LITTLE BOATS AND BIG SALMON

By Erv Jensen

Little Boats and Big Salmon

First Edition

Library of Congress Catalog Card Number 98-90980
ISBN 0-9668753-0-3

Published by
Three Trees Publishing
P.O. Box 333
Silverdale, WA 98383

Telephones (360)692-2117, (360)698-3533, and (360)698-3907

Printed in the United States of America
By Gorham Printing, Rochester, WA 98579

Cover by Kathy Campbell, Gorham Printing

Copies of this book
are available from the publisher.
Find Order Form inside back cover.

To Donna, Valerie, and Joy

CONTENTS

INTRODUCTION

In all the years it took me to bring forth this book, never did I doubt the clout of the subject matter. Battling big salmon from little open boats among the remote, rainforested waterways of panhandle Alaska – here was fare to stir the throwback soul. Doing it with light tackle added another dimension. Turning recreation into livelihood via commercial fishing license was no stunt to shun, either.

Some of my enthusiasm found its way into columns which, back in 1956, I began dispatching from the field to my hometown newspaper in Bremerton, Washington, *The Sun*. Fishing was already popular, and this was *Alaska* fishing. The readers were duly impressed, and said so, in gathering places around town and in letters to the editor and to me. "You've got it made," one told me "You get paid to catch fish and paid again to tell about it. I think most of America would like to have your job."

Enhancing both the fishing adventures and my chronicling of them was my brother Svend, the ideal partner: fan of the wilds, critic of nothing, and student of our fishing technique, *mooching.** For his growing expertise in this art of feel-fishing a bait into the lie-holes of lunkers I quoted him without restraint, and he became mildly famed. I dubbed him "Brother S" and, across *Sun* territory (circulation then about 25,000), he became "Brother S" to folks not even cousins.

So what held up the book? Too much material, a problem which could only grow worse by the year. Once I lost a 50-pound salmon in my files. I knew I had entered him, along with the details of his capture, but where was he? Not under TROUBLE, although he had given me plenty of that, nor BIG FISH, where he certainly was eligible. Finally, remembering the buckshot rainstorm that assaulted me during the battle, I located that great fish under – WEATHER! I apologize to him.

**For the uninitiated, mooching is an intriguing descendant of trolling. If while trolling you slow down and let your bait sink to the bottom and then pull it up a ways or otherwise make it behave erratically, meanwhile keeping your senses tuned to the feel of any nibble on it – you are mooching. Invent your own variations.*

The headings TROUBLE, BIG FISH, and WEATHER all eventually gave way to others. The fish stayed in. You'll meet that one under WET DAYS, starting on page 30. Others slipped from top billing. Consider the problem. If a certain gem of acquired lore produces a 35-pound chinook in 1979 and a 45-pounder in 1999, the 35-pounder becomes expendable.

Brother S (left) and Brother E on a cold Alaska day long ago.

Brother S and I became fishermen in the 1920s, moochers in the 1940s, and commercial Alaska moochers in the '50s. So does this book lament for good old days? No. Southeast Alaska (or plain "Southeast,"as it is optionally called), is still here, still healthy, and still populated by scarcely more than one person for each of its 40,000 square miles. It is still raw and robust salmon country. It escaped the Big Oil Spill of 1989, being separated by the Gulf of Alaska from that tragedy in Prince William Sound. What we do in this book can still be done.

Indeed we have seen changes in our years of Southeast mooching. The seasons have been shortened, the restrictions tightened. Too bad? Not entirely. The time you can't spend enhances what you can. Grab your rods, reels, and rain gear.

* * * * * * *

Thrills, woes, risks – in an Alaska mooching career you take them all as they come. But there's one thing you'll never do more than once a season: catch your first fish.

"All first fish are good," Brother S will tell you, "but some are better than others. And of these, one is bound to beat all the rest. You'll remember funny little details."

Mid-May, 1978. The day was robust – "unsettled," to the weather bureau. A shower had moved on and now blue sky showed in patches. Sunshine sparkled in the curling bow wave and flat wake of the 20-foot ragtop Sabercraft we were using together that year. Astern, between the green hillsides of Revillagigedo Island on the south and Betton Island on the north, Clover Passage narrowed away as we planed ahead.

Brother S suddenly doused the drone of the 85-horse Merc. With a long *whooosh* we settled in our wake in Tongass Narrows.

"Can't make it," he said. "Have to let her go first."

"Her" was the big, blue Alaska Marine Highways ferry *Taku*, northbound out of Seattle, Prince Rupert, and Ketchikan through the forests for Wrangell, Petersburg, Sitka, Juneau, Haines, and Skagway. Against the sky at the rail atop her, a dozen passengers stood abreast, swiveling heads at one another and pointing down at us. I stood up and waved, and at least 10 of the dozen excitedly waved back, obviously pleased at being noticed. Brother S chuckled.

"They've come to Alaska, so they're Inside, as the saying goes. But we're more Inside than they are."

He goosed the Merc and we rode over the *Taku*'s waves, back up on plane.

Off the Vallenar Rocks ahead, sea gulls were wheeling and plunging. S leaned into my ear.

"They're right over our old hot spot. Must be after herring – chased bait!"

Chased bait! The words unleash adrenaline. A picture flashes: underwater violence. Herring driven to the surface by attacking salmon. Ah! For a fisherman with a crippled-herring mooching bait, no salmon is more catchable. And these salmon would be kings, chinooks, big, brawny Alaska chinooks, bright ones, early in the spawning run.

The gulls were doing their gyrating precisely over an old "fishy triangle" of ours, a triangle of which side *A* is the outermost Vallenar Rock, side *B* a border of kelp, and side *C* a tiderip line. Rocks, kelp, tiderips. Great places for kings. Of your first thousand fishing holes, 900 will claim one or more of these attractions. This one has them all....

Brother S cut the power, and the silence of the realm descended. Civilization was far away. The gulls had settled on the rock. Were we too late? I tilted the 7.5-horse kicker down and fired off. Its soft exhaust at mooching speed was pleasant company in the wide world. Fishing time!

With a good partner you may work in silence. We often do. While I run the kicker Brother S cuts the bait and dresses the catch. Fast. From a 10-inch herring peeled from a thawing block of so-called "halibut bait" he can slice off two deadly spinners in 15 seconds.

We use 9- to 11-foot rods, Penn 109 reels, 20-pound nylon lines, 1- to 4-ounce sinkers, and, for chinooks, 3/0 hooks. One of our systems when we fish from one boat is to use two outfits each, working one and trailing the other, close to the boat. We hung out the short lines and were about to sink the long ones.

Zinnnnngg! One year it took us three days to hook the first chinook. This one grabbed S' short line in three seconds. The rod whipped, the reelspool shrieked, and Brother S scrambled.

"King!" he wheezed at me. To the fish he said, "Stay away from that kelp!" The fish took heed – ran away from the kelp instead of into it. The fisherman, in charge if shaky, prevailed. In 15 minutes I christened the net and splashed the prize down on the floorboard.

"Beautiful ... fish!" whispered S. With a knobby old thumb he lifted the gill plate, saw a tinge of crimson in the throat latch, and added, "and a Red at that!" Why some kings are red-fleshed and others white is an old question. Some modern biologists, seemingly supported by the results of controlled feeding in fish farms, say diet does it. Many fishermen remain unconvinced. "Red or white, they're all eating the same herring," says S. What is not debated is that Reds bring the higher prices, and Large Reds the highest. Large Reds are from 11 pounds up.

This Red weighed 37. It was the largest first fish ever, besides being the quickest to lay hold. But the fish box remained unopened. "Can't put him in yet," said Brother S, straight-faced. "Have to keep him out to look at for awhile first."

The looking takes about 15 minutes, off and on, even after all our years. That's the way it is with first fish.

1.THE SEVEN SECRETS OF SOUTHEAST

In the year 2030, after you've been piling up experiences in the Southeast mooching life for 30-plus years, you will be moved to write your memoirs. The world will have become ever more structured and scientific by then, and you will feel obliged to be systematic and analytic in approach. You'll ponder and muse. Just what are the factors that make this life magical?

As of now I'll say they are these:

The Instant Outback

A man was mounting the marble steps of the Federal Building in Juneau with a shotgun on his back, and I, being then a cub reporter new in Alaska, enthusiastically recorded this "How about that!" scene with the camera I carried. This was in 1952, but the scenario could easily be the same today. When I asked the man for information for a caption he was cooperative, but surprised at my interest. He was merely "going ptarmigan hunting up Mount Roberts and mailing this letter on the way."

My editor at the daily *Empire* sided with the ptarmigan hunter – no big deal. I, he noted, had come from Kitsap County in the State of Washington, not far from Seattle. "You are in Alaska now," he told me. I was tempted to reply, "And you've been here too long," but I liked my job otherwise, the salmon fishing being good around Juneau, and I managed silence.

In ensuing months, subjects not space-worthy included an eagle perched on a downtown hotel railing and a black bear impeding traffic on Glacier Highway. A blackie that invaded the city did make print, although the emphasis was on a recipe for stewing him.

On balance, then and up through now, appreciation of wilds-outside-the-windows has remained demonstrably strong in Southeast. We Jensen brothers encountered one example, and some neighborliness besides when, out exploring, we trolled up on a neat little cabin set against an endless stretch of the woods. We knew not whose. But the unlocked door said "Welcome," and the walls inside bore paintings of wilderness living. Brother S went back outside. "Beautiful pictures," he said. "But why does he need them when he's got this?"

On another trip we entered a remote bay. Nobody here. We owned the place and everything in it. "Everything" proved to be plenty: brawny chinooks, big and hungry, waiting to tangle with us. We obliged. The skirmishes ended with three soakers, totaling 107 pounds dressed, iced down in the fishbox. Then came the contemplative pause.

Clarence Strait lay empty before us and the forest rolled away endlessly behind, stretching northeast beyond Alaska into the vast reaches of the Yukon and Northwest Territories.

"We're a million miles from civilization," I said. "But we're still only half an hour from Ketchikan."

"And besides that," S said, "Ketchikan is a metropolis with suburbs. Out of smaller settlements you could get lost in two minutes."

A routine day's scenario may unfold about like this: Where a few small buildings sit square against the forest and a float extends into a cove, you and any compatriots board your boats in the morning, exchange "Good Luck" wishes, and disappear around favorite bends, not likely to meet again until you come back out of the woods at dusk.

Meanwhile there is your "work." A salmon of 50 pounds or more on a 20-pound line gives you plenty of it.

Since you are a commercial fisherman with several lines, additional soakers may add to the workload. For this reason among others, it is nice to have a partner, either in your boat or a separate one. Not that he'll do your work. But nobody, after spending a day like this, alone with the trees and the fish, should risk having no one to share the highlights with.

OUTBACK'S EDGE:
Here's Knudson Cove, Alaska, our fishing base. It is also a link between civilization and the wilds. To the south, the city of Ketchikan is 12 miles away by highway. But to the north, beyond the trees in the background, you could go across the rest of Alaska and across the Yukon and Northwest Territories of Canada and over the tundra to the Árctic Ocean, and never see another human being. Photo taken 1960.

Custom Quarters for Fish

Rocks, cliffs and trees eased past our starboard gunwale as we putt-putted quietly along, gear in the water. The tide was at a substantial minus, revealing shoreline features not usually up for inspection. In the face of one sheer cliff, the water coughed and sputtered at the level of the wash. Brother S stood up, leaning and peering. "There's a cave down there," he said....

We swung closer, and could see a substantial black maw. "I wonder," I said... "Would salmon hide in there, like trout under a log or beneath a bank in a stream.?"

"We'll have to come back and try it when the tide's higher," he said.

By reference to a blotch of darker rock beneath the trees atop the cliff we marked the spot. We were not greatly excited. While we would indeed be back, this contour was still only one of the thousands that enhance the islands of the Alexander Archipelago.

MOOCHERS' REALM: Here's a typical scene, both of our workplace and of Brother S at work in it.

The islands number 3000 to 11,000, depending on who is defining "island," a belt of them 100 miles wide and 300 miles long, strung out in all shapes and sizes along the mountainous mainland. Fascinating, they are, these islands, but the real treat if you're a fisherman is the spaces between – channels and passes, guts and bores, wide places and narrow, shallow and deep; reefs, bights, bays, coves, arms, eddies, backwashes. Tumbled bottoms. Salmon haunts. As Brother S put it the first season we came here, "This place was made for mooching."

The day proceeded approximately thus:

10 a.m. Mooched along a cliff so steep it appears to lean seaward. We come here not so much for fish as for weird sensation – tons of rock hanging overhead.... Once during a blowing rainstorm Brother S suggested we "give it a try," but I was reluctant. "We never catch anything there," I said. "No, but maybe we'll be out of the rain," he said. So we went, and caught nothing.

12 noon. Squeezed between two spires, nearly brushing the vines hanging from one, while probing into deep water below. Fifty feet further along, the bottom so rose that three kelp reached from it to the surface. Alongside the kelp a 31-pound chinook laid on.

2 p.m. Wove a crooked, careful course among an assortment of rocks, some awash and some high and dry, perched among places 100 feet and more deep. Hooked a 20-pounder and pursued him in this obstacle course, relieved that he wasn't bigger. Once, a minor swell lowered the prop noisily on a rock. No damage. Bounced the net off another rock to land the fish.

3:30 p.m. From a stretch of Clarence Strait known for the fury of Southeasters, sneaked through a gap in high rocks into a "God's Pocket," a gymnasium-sized retreat with a smooth gravel beach beneath the evergreens, and sat a spell with cheese sandwiches. The taste of the cheese grew mixed with a leathery odor in the air. "Bear," said S, an old-time hunter as well as fisherman. "Some of them don't smell so good." We had seen a blackie in the area previously, but he did not show himself now.

5 p.m. Trolled for half an hour over a soft, level bottom, an underwater prairie that was an extension of a smooth, sandy beach, one of the few in our fishing territory. Brought up a 20-pound halibut, but halibut were out of season. Sent him back to his underwater flat.

7 p.m. Returned to the cave. Made a pass within 20 feet and drew a blank. Tried another 200 feet out and hauled in a 28-pound chinook; worked close in again, and here, on my deep gear, got a four-foot, 50-pound ling cod. In the net he gave us a bug-eyed, mammoth-mouthed, spike-fanged, displeased glare, but we didn't blame him any and, in as friendly a way as we could, considering his heft, dumped him overboard again.

9 p.m. Eased up to a waterfall plunging 50 feet directly into salt water. Here you could fill your fresh-water tanks without going ashore, although the spray would get you a bit wet. Elsewhere, streams run quietly down rock faces, and refilling is simpler.

NEED FRESH WATER?
Ease up to the falls.

Havens for Fishermen

The chinook goes about 25 pounds. You ease the fishbox lid down on him while thoughtfully swinging your eye around this secret place where you caught him. It is a ballfield-sized retreat. The tide is out, and the cleft-rock banks slope up like bleacher seats – no spectators in them. You are the only person in the world privileged to be here.

The water is deep, the surface so flat you could have landed your fish standing up in a kayak. The evergreens loom silently all around, crowding the shores and increasing the concealment. You can no longer even see the narrow pass through which, lured by the explorer in you and a few herring flipping, you slipped in.

You have done more here than catch a beautiful fish. You have filed a sample of topographic hospitality. If the underwater parts of Southeast are contoured to invite fish, the parts above are shaped to accommodate fishermen, offering both escape from competitors and shelter from storms....

And another time: A 30-knot Southeaster was barreling up Clarence Strait, and the Jensen brothers retreated into a narrow-necked inlet. The surface here was still too rough for comfort in the 20-footer we were using together that year, and we went around yet another bend. Here we foiled the blow almost entirely. While the spruces above waved furiously, mere puffs of air rippled our millpond and cooled our faces.

And the kings cooperated. Before the Southeaster subsided and we emerged to work in more conventional – and in fact more likely – places nearer main migration routes in the strait, we had boated two.

Some of Southeast's islands bend into horseshoes around others and some jigsaw loosely together, separated by waters not safely navigable in boats exceeding mooching size. As S said during one of our winding journeys of exploration, "Within five miles of where we are right now, there's 50 miles of shoreline."

In my first travels into Southeast's intricacies in 1953 I nosed my 16-foot plywood runabout ashore on one of the few gravel beaches to be found in this rocky realm. I was joined there by another fisherman, the only one I had seen at close range all week. We talked awhile.

"Spend a lot of time around here," he said. "Been doing it for 20 years. Fish in the summer and trap in the winter."

I asked him what was on the other side of our island.

"Darned if I know," he said. "Never go that way."

I had heard right. Twenty years, and never been there! Then were there places here where no one ever went?

I trolled around the island – and found only the other side. The next day I went beyond it to the next one, and again found nothing spectacularly new. In a rainstorm I went ashore, shook out the kinks, and ate a sandwich hunkered under a dripping hemlock. More or less uncomfortable, I got compensated while leaving by catching a chinook a few feet off the beach.

Three years later Brother S and I took this same route, again in open boats and again in the rain. This time we lunched in a dry shelter under an overhanging rock a mere 200 yards beyond my dripping hemlock.

You can get frustrated pondering the places you haven't been. "We could go scouting by air," I suggested to S once. He grinned. "That'd spoil the fun. Using charts is bad enough."

DIVERSE PLACES: Shores where you can land a boat comfortably are not numerous in our Southeast Alaska fishing territory. Above is a nice one; below a more typical example.

Southeast's settlements of Ketchikan, Wrangell, Petersburg, Sitka, Douglas, and numerous lesser ones all are on islands. Juneau is on the mainland but still accessible only by water and air, nearly 100 island miles inland from the open Pacific. Islands everywhere. In one lifetime you will never get acquainted with them all. Good. There will always be somewhere new to go.

Governmental Blessing

On a June afternoon in 1982 we pulled into Knudson Cove and, in a row on the public float, laid out six sizable chinooks for transport by pickup to the fish buyer in Ketchikan. This operation, as it frequently did (and in the 1990s still does) attracted a few curious tourists.

"I think we've been paying about $4 a pound for salmon in Minneapolis," one serious chap told us. "Your work must be very profitable... and enjoyable."

"You catch fish like these with gear like that?" put in another, eyeing our light rods in their racks atop the cabin. "Profitable? What's amazing to me is that it's possible."

On the way to town Brother S reviewed the appraisals. "Profitable, possible.... What they missed is that it's legal."

Consider a bit of history.

Fishermen on our native Puget Sound in Washington State either originated mooching or were quick to take it to heart. They were enthusiastically into it at least as early as 1940. To some enterprisers among them, the effective and fascinating technique offered opportunity beyond sport. They bought commercial licenses and got paid for recreation.

They called themselves kelpers and their work kelping, labels based on their small-boat, sport-tackle maneuverability in kelp-laden, rock-strewn territory. These were waters untouched by power trollers, and the kelpers prospered, until –

The state, faced with a problem arising in a barely related area of regulation, reacted by outlawing kelping. Organized kelpers fought the order and beat it in court. But the state revised its approach and tried again, and this time prevailed. Hand-held poles were outlawed. "Fixed gear" was required.

"So I bolted my nine-foot glass rods to my gun'ls, and that made me legal," said one disenfranchised kelper. "It sure didn't make me happy, though."...

Now savor a contrast. When I arrived in Juneau in 1952 to pursue dual professions of newspapering and commercial mooching and took my first catch to the buyer – bingo! Rod-and-reel commercial fishing was not merely tolerated; it was officially blessed, in print. The method-of-catch boxes on fish-buyers' tickets included one specifically labeled "sport fish."

The explanation was hardly that Alaska had no regulatory problems. The administrative history of the Great Land's commercial salmon fishery was already a history of stress, as I learned on my other job as a newspaperman from Clarence Rhode, then the top administrator as regional director of the U. S. Fish and Wildlife Service.

"The wildlife part of this job – it's easy," he told me. "But the fisheries? A big headache. Too many fish traps. Too much other gear. Too much argument – too many gear fights. Everybody wants everything."

The traps, except for those of Native tribes, were outlawed soon afterward, but the register of competitors today remains formidable.

Atop the list, mass-haul-wise, are the purse seiners. They may dump thousands of fish aboard in a single set. Following them, in order of decreasing gear heft and increasing attention to individual fish, are the gillnetters, power trollers, and handtrollers.

Within the handtrollers (the "sport fish" box of the fish ticket has been absorbed into this category), the descent continues. There are the hand-gurdy, wire-line types, the regular rod-and-reelers, and finally us ultimate lightweights, the moochers, the most tuned-in of all. To us, every fish is a worthy and exciting individual.

And we get him with the blessing of the regulators.

We also have been getting – let's say in recognition of status – more attention from the regulators. One rule that came along, for example, forbade fishing with hand gurdies and sport rods at the same time, and another banned the use of a heavily-weighted separate line or "downrigger" to take the sport gear deep. We lightweights must be lightweights unadulterated. It is not a burdensome restriction.

Says Brother S: "We can't catch enough fish to hurt the resource. But we can sometimes make a little money, and always be out here taking in the country."

Staying Power

At 5 o'clock in the afternoon on June 23, 1978, having completed a day requiring my presence in Kitsap County, Washington, I was part of the roar of wheeled humanity on Washington Interstate 5, bound for Seattle-Tacoma Airport. That same night, from a secluded cove among Southeast spruces, I mooched up two king salmon totaling 82 pounds – 69 pounds dressed. To enable this neat trick, Brother S merely met me with the boat in Tongass Narrows alongside Ketchikan International Airport.

The airplane ticket cost $77.50. The two chinooks sold for $141.45.

You could come up with various conclusions as to what we had proved, including the "instant outback" and "fishermen's havens" ideas of previous pages herein. The main motive in fact was to get back to fishing. But the feat did merit entry into our treasury of memories, there to stir up others, including one filed on an evening in June 1955.

We were then still new refugees from Washington State, still in the flush of realization that Southeast salmon and the Southeast mooching life were real things on our earth. One kind of proof had come from an old troller, who cautioned us to "pay attention to where you are" in the snarl of islands and wooded waterways – "you could go out some day and get lost." Other bits of reality included our tent, snuggled among the huckleberries and skunk cabbages of Knudson Cove, and our two 16-foot outboards, moored at the one small float in the harbor.

In these two craft we had just returned from Vallenar Point, 15 minutes away, where in the total absence of other fishermen we had landed four 30- to 35-pound kings in the last hour of daylight – a scenario unlikely where we came from.

In the tent our emotions were mixed. "We got here 20 years too late," I said. "This can never last."

"Well, at least we are here now," S said. "In another 20 years, these will be the good old days. And we got in on them."

Twenty years, we had said. But in 1978, not 20 but 27 years later, the episode of the two "air fare chinooks" had given us a lesson in prophecy and a new file for the treasury. Tidbits fitting that file have been presenting themselves ever since. In Southeast, in other words, the good old days still are with us.

Back in the 1960s a Seattle columnist wrote that the world of the immediate post-World-War-II period "has totally passed away."

Since then, the world as of then, with its own modernities – dollar-a-gallon-plus gas, computers, space shuttles, lasers, and fiber-optics also has grown old.

But Southeast remains Southeast, stubbornly resistant to domestication. Populous Anchorage is two mountain ranges and the Gulf of Alaska away. So is the oil pipeline. To an editor in Anchorage, the Panhandle is "a dead area." But to one old-time government administrator who worked across the Great Land for 30 years, "Southeast just hasn't been ruined. It is still the old Alaska."

Said another Panhandle booster:

"The population hasn't zoomed, and wouldn't hurt much if it did. Say the area of Southeast is 40,000 square miles and the populated places cover 10 square miles. If you double the population and the area populated, so that you now have 20 populated square miles, how many do you have still unpopulated? 39,980 instead of 39,990."...

All this is scarcely to say that in four decades we have seen no changes. The chinooks are fewer, the restrictions tighter, the seasons shorter, the competition stiffer.

Except among incorrigible purists like us, mooching itself seems to have slipped as a favorite technique. Many of the calm, close-in waters – those ideal for mooching – have been closed to commercial fishing. Some past moochers have succumbed to improvements in artificial lures and hardware. New moochers are few. Alaskans seem reluctant to depart from 40- to 60-pound lines and 12-ounce sinkers.

The positives still predominate. Equipment is better – boats, motors, communications, electronics. What has been lost does not match what remains: seclusion, salmon, scenery, rain, deer, eagles, porpoises, whales, wilderness characters. The instant outback. The secret places. Don't fear that any of these aren't still here, nor either ever imagine that you couldn't still, as the man said, get lost. Southeast is still Southeast.

Bonus Benefits

Tenderly, Brother S rubbed the shinbone which, minutes before, he had barked on a fishbox corner while jockeying for position with an outsize Southeast coho. "Why is it," he said, "that you never catch a fish without something extra to remember him by?"

It wasn't a question to be answered, only appreciated: always something extra.

The Fisherman's Extras that bloomed in Alaska for me began when I caught my first keeper trout at age 5, 75-plus years ago.

He measured seven and one-half inches, but otherwise I do not see him clearly. What *is* vivid is that I took him from under a mossy bank at the bend of the gurgling brook in the woods of our farm on Puget Sound, with a cut willow pole, on a worm, and, yes, with a bent pin for a hook. That pin – how I acquired it so I could bend it is a further Extra. When my mother's sewing box yielded none, I cleverly snitched one from a dress she was hemming. Red print dress. All those years ago.

Extras. The Extras of that episode, and those of the years since, tell us these things about them:

– They need not be important. They have their own reasons for demanding immortality, and they do not share these with you. The memory persists and you don't know why.

– They are of infinite variety – Extras of joy, excitement, discomfort, fun, risk, frustration.

Extras may be either good or bad, though most bad ones are less bad in memory than in the making. The fishhook that a 49-pound, 9-ounce king rammed into your right index finger at the first joint produced the staggering pain of injected fish slime at the time of injection, but the residual scar is a treasure.

Once I caught five great kings in the last two hours of daylight.... But did you ever dress five aromatic kings near a Southeast shoreline in a June dusk? The gnats, chiggers, mosquitoes, and no-see-ums are then and there their fiercest. Slap them? Your hands are dripping fish slime.

Extras often stem from fear of losing the fish. What sticks with you is the snag, the backlash, the reel spool shrinking bare.

Once upon a time the newspaper where I was toiling got word of the catch of a monumental 68-pound salmon. When our photographer caught up with the lucky fisherman, he was prosperously puffing a new cigar, and for his picture he clamped down mightily on this while muscling his monster aloft.

But his quote stressed not fish size but excruciating suspense – "Almost lost him. Had to untangle him from three other fishermen and chase him clear around a kelp bed."

The picture went nationwide by wire, and the next day a frantic call came from an east-coast advertising agency.

"Are you there? Ah! Contact! That picture! We want to buy that picture! Have to have an original print!"

"Well, it can be arranged... You selling fishing gear?"

"Oh, not fishing gear. Cigars."

Ranked high among Extras are those attending extended sessions away from base camp. You forget the salt and – inspiration! – boil the potatoes in sea water (we did, and it works). You find your bait going rank, use it anyway, and knock the salmon dead (we did that, too).

Extras prefer to crop up unanticipated, and you young fishermen generally must take them as they come. Oldsters learn that Extras can be planned or promoted – rigged, say. Many are the sessions we have spent on a remote beach on Prince of Wales Island where tidal currents deposit the flotsam of far places. In the fascinating expanse of baubles and junk we rummage an hour while the fishing waits. Rubber balls, rum bottles, light bulbs, dead ducks, citrus fruits, fishing plugs, glass floats. Lifejackets, boat parts, clothing, comic books, furniture. One year Brother S extracted a wooden rocking chair from a tangle of miscellany at the base of a big stranded stump. He set it atop the stump and perched in it there, surveying his domain.

Another time we nosed our outboards up a forest creek at high tide – and combined a rigged Extra with a chance one. Our strategy was to sit a spell in a boat surrounded by timber, an adequate Extra in itself, even if rigged. What was not rigged was the black bear that rose from the grass beneath the gunwale, sneezed, snorted, and fled splashing up the creek. We went the other direction, back to our offshore fishing.

To appreciate the full range of your own Extras, review your collection after a few years. But be prepared for confusion. Are most fish special because they come with Extras, or most Extras special because they come with fish?

Super Humans

Characters. Legends. Handlogger Jackson, Buckshot Woollery, Crazy Al. Scattered recluses. One of these may meet you with a .375 Magnum when you enter his lonely bight – and detain you with endless friendly philosophy once he deems you acceptable. Many others. Not the least of the elements spicing your Southeast mooching life will be – people. People tuned to the country. People of unusual vocation or avocation, some of them important, some only interesting. Special people. Look forward to them. We look back on these:

– Robert T. Baade (pronounced body), government fisheries biologist, whose professional work required him to sport-fish for trout – to travel by plane, boat, and foot to Southeast's lakes and streams and to determine and record what prizes sports fishermen might catch there and how best to catch them.

– Walter Sperl, water-borne postman who, on many a black and frigid night on his lonely routes across Southeast in his 50-foot mailcarrier *Yakobi,* chopped ice from her rigging in flying seas to forestall perilous topheaviness and keep the faith: the mail must go through.

ICED UP: After a frigid, stormy night carrying the mail across Southeast, Walt Sperl's 50-foot Yakobi *often came back to Juneau looking like this, and sometimes worse.*

– Dorothy MacKenzie, Ketchikan businesswoman arising before dawn so as to spend hours on the water manning an outboard, cutting bait, and catching salmon before opening her ladies' fashion shop in town for the day.

– Mrs. Ed Glenz, postmaster and storekeeper at Myers Chuck when, in 1964, Brother S and I stopped for gas at this forest-squeezed, island-fronted village – a place to be found only if you know it's there – 40 remote miles up Clarence Strait from Ketchikan. Her space included a garage-sized building with a shed alongside. Her mail and general-merchandise patrons she served from the main building, but for some customers she needed the shed because, she said, "you can't sell beer in a post office."

– A disinterested Myers Chuck fisherman named Kennedy, whom we talked to on the float shortly after our visit with Mrs. Glenz and who told us, "I may go out pretty soon. It must be 8 or 10 o'clock, isn't it?"

These doers of deeds ordinary in Alaska:

– A Juneau man who, noting that I was new in town, invited me on a drive out Glacier Highway to the small boat floats at Tee Harbor. There he took the line of an incoming boatman and made it fast to a cleat. They exchanged pleasantries and the man went ashore. "Seemed a likable sort," I said. "Who was he?" "Don't know," my friend said. "Haven't seen him before."

– Another stranger who spotted me momentarily dead in the water halfway through a 12-mile crossing of Clarence Strait and veered half a mile off course to swish to a stop alongside. "Trouble?" he asked. "No, just changing tanks," I said. "O.K.," he said, and went on his way.

– An Ellis Airlines pilot who treated his passengers to an unscheduled stop by landing his Grumman Goose near Brother S' boat off the Vallenar Rocks and telling him, "There's a boat adrift down in the pass. Needs help. You'll be able to see him as soon as you run down a ways." Brother S found the craft – an inboard-outboard which had cracked its lower unit on a rock – and was towing it to Ketchikan when the Coast Guard arrived and took over.

– Gov. Jay Hammond who, in December 1977, on overhearing a radio message while flying his private plane, turned abruptly off on a side trip to land at the scene of a crash on a frozen lake and fly out three survivors.

It may be well to acknowledge for balance that in Southeast you will also find phonies, deadbeats, oddballs, misfits, and rebels, critics of the establishment. During the heydays of the hippies a few decades ago, Ketchikan surely had the dirtiest, crummiest, hippiest hippies loose anywhere on earth.

But then, you heard no claim in the beginning that all the people you'll meet will be heroic or high-grade. Most seem to be, and you'll be thankful for that.

WORTH THE WAIT: When the tide comes in, the boat above can float into the "God's Pocket" below, where Brothers E (left) and S are enjoying a break and a fire.

FINISHING UP: The dressing job completed, Brother S rinses off another chinook.

2. THE UNRULY ATMOSPHERE

Southeast Alaska was created under a general principle calling for the avoidance of moderation. If you have come here from the Puget Sound country of Washington State, as substantial numbers of us have done, you'll find the consequences to include the enhancement of the natural features you enjoyed in the more rugged of the fishing grounds there. The forested hills are steeper, the rock cliffs above the salmon haunts higher, the salmon mostly bigger. The tides rise and fall further, requiring that man's waterfront structures stand higher on longer piling. And so forth. And then there is the weather....

Wet Days

As a Southeast fisherman you will learn to like rain. Encouragement is available from resident enthusiasts. Early help for me came from high-placed federal officials on my news beat in Juneau, in northern Southeast.

B. Frank Heintzlman, regional forester (later governor), the man in charge of Alaska's billions of trees, smiled out in professional approval at the dripping spruces. "One thing we don't have to worry about in Southeast is forest fires," he said.

Clarence Rhode, the U. S. Fish and Wildlife Service's top officer in Alaska (regional director), told me, on the first warm afternoon in June, "When you get a summer like this you don't mind the other 364 days." He added, "I didn't originate that remark, but it's mostly true. Too much sunshine here irritates people. They want it to rain again."

C. Howard Baltzo, Rhode's assistant, later to be named director of the Pribilof fur seal program, spoke of "joining what you can't beat." He invited me to "come along on our annual cruise and picnic next Saturday." The rain was wrestling on his office windows.

"Weather permitting?" I asked.

"Weather notwithstanding," he said. "If we waited for weather we'd never do anything."

So the picnic proceeded, the picnickers toting slickers and Sou'westers. It didn't rain. Too bad.

A fisherman with whom I stood in a downpour at the Tee Harbor floats outside Juneau said, "You think this is rain? Wait'll you see a buckshot rain."

Later in southern Southeast I learned it doesn't really rain around Juneau. Precipitation there is about 55 inches a year. In Ketchikan it's more like 150. One year, 1949, it passed 202.

The only thing more memorable than fishing in a southern Southeast buckshot rain is battling a lunker fish in it, alone in a 16-foot open boat. In newsprint I said this:

> For four days straight the sun had shone. Equalization was inevitable. As if enraged by postponement, the gray black clouds came rumbling down over the spruces on the hills. Like a division of tanks they assembled for bombardment.

The water, flat despite some wind in the dark dampness close above, assumed a look of night: roiling, black, ominous. I drank in this ugly beauty, knowing the barrage was coming.

The temperature plummeted. Shivering, I dutifully checked and adjusted my baits, put on my mackinaw under my raincoat, donned a pair of fleece-lined, cork-and-rubber-coated fish-handling gloves and, squirming deep into all my garb, froze, and waited.

The tanks rumbled closer, then kicked up their snouts. That was the recoil. The buckshot had been fired. In seconds it would hit.

The king hit first, barely. He took my short line full bore, streaking horizontally. Then the buckshot hit, smashing the surface with a sustained roar that raised a million geysers.

My salmon fled the barrage at 45 degrees down. The reelspool was a blur in the deluge. My hands were in boxing gloves. The fish swam. The buckshot pounded, roared, and splattered. I made helpless motions. Floodwater stuck the gloves to my hands as a boot sticks in a mudflat. The king kept taking line.

RAIN? One year the total exceeded 202 inches, per this sign that stood for years in downtown Ketchikan.

With my teeth I tore viciously at the right-hand glove – I bit a hole in it, broke the suction, and got it off. The buckshot moderated and the fish came up. Twenty minutes later I splashed him down inside. Fifty pounds, two ounces....

Ketchikan, the "salmon capital of the world," is also the rubber boot capital. You find bankers in boots, physicians in boots, lawyers and bartenders in boots, women and children in boots. The air says boots. When it stops raining, it is preparing to rain.

From the wet hemlocks and wet spruces on the wet hillsides, the wet feel washes down on you. It is substantial, robust, lusty, like a breath of fresh wet air. Vapor, condensation, dripping equipment, sogginess. Wet hands, wet boat cushions, wet floorboards, wet fish. The wet is wonderful.

The rains come in vast varieties of three major types, only one of which will get you wet if you are appropriately garbed. This is the horizontal rain, a windblown jumbo mist that seeks out your face and runs down your neck.

The second type is the standard rain, moderate, soothing, satisfying.

And buckshot rain. Several inches of it may roar down in an hour. From the creeks the water spews in walls, and the coves at the mouths take on the brown, freshwater-look of lakes. You could probably drink the surface stuff. Your oilskins shed buckshot all right but it still adds pounds to your weight, by pushing. Staying dry in all that water makes you quiver with delight. All you need for the ultimate memoir then is a big king pulling at the same time.

Windy Days

In your early letters back home you'll aim to straighten up those misinformed old friends who think you must be constantly battling mean weather and high seas in wild Alaska. The open Pacific, you'll explain, is 100 miles away. Its fierce storms batter themselves flat on the thousands of islands between. You are pulling those cantankerous soakers out of quiet water.

Your next problem will be to explain how the wind got through to blow you off the water again today.

One season we couldn't tune in on storm warnings because the radio tower blew down. Shipwrecks? Only one (see page 87) and a few close calls.

"Those people who think a fast boat will get you out when a storm starts – they're dreamers," says S. "It won't. You've got 10 minutes of open water to cross and you're in high seas in five. Anyway, how do you guess whether one puff of air means anything?... especially if the fish have just started biting?"

Go prepared. In a tumbled old cabin overgrown with salmonberry bushes in Gambier Bay we found evidence of someone who didn't. The dirt floor in front of the rusty stove was piled high with mussel shells. The scene presented itself in the gloom: a stormbound hunter or fisherman subsisting on broiled mussel on the half-shell. The pile was two feet high. "That storm must have blown a week," S said.

The winds became the subject of a discussion we were having with a visiting sports fisherman from Chicago, and Brother S gave him a rundown on variety: "... some monsoon types, some chinooks, some williwaws, and some crazy ones that reach down and dig chunks out of the woods." He aimed a finger across the water at Cleveland Peninsula. "There's two holes in the trees over there – two I know of – that got there just that way."

Along the shores we fish most, those of Clarence Strait from Chasina Point north to Ship Island, the prevailing winds are Southeasters. Fish-rich Caamano Point sticks its rocky snout straight into the oncoming wind and water. Hang a fish while alone in a boat in a place like this and you need four hands. Rod and motor require one each, and reel and net two more. You pitch, plunge, and sway, while the fish and your sense of balance dicker for your attention. Ashore, frothing at the maw, the rocks wait.

One morning in the hushed calm following a big blow we arrived at the Caamano shore while being overtaken by a Coast Guard helicopter. In another boat a man was waving the chopper into a bouldered cove piled with shredded logs and drift. Projecting stern first from the mess was a smashed aluminum skiff. Other guardsmen arrived by water and poked among the rubble. They found nothing, and we never heard any news of what had happened.

That same afternoon Brother S got into difficulty in a mere breeze. In separate 20-foot boats that year, we were "working No. 2 Reef behind No. 1 Rock." The reef runs from the rock toward shore. Fishing anywhere nearby is "working No. 2 Reef," but S does favor the reef itself – "To go through there the fish have to come up, so they always see the bait."... A chinook saw it, took it, and ran. Brother S, following, lost the protection of the rock. Wind and chop shoved him shoreward.

The salmon dove into a deep spot and stayed there. With motor idling in neutral, S drifted over him, applying pressure. Standoff. The boat bobbed toward the rocks. He put the motor back into gear. Thump! The prop hit something. The motor died.

Rod aloft, he yanked the starting cord with his free hand. No luck. He jammed the rod into a holder and yanked with both hands. No luck. His boat was en route to join the aluminum job on the logjam. We have a rule: never shut off one motor before starting the other. This violation was unintentional! While I was debating whether the situation warranted some kind of assistance from me, S dove into the cabin and turned the key – and the running motor took right off. He circled outside and, shuttling between cabin and cockpit, landed a 38-pounder.

The underwater obstruction proved to be a sunken log. The trolling motor ran again on the first post-emergency pull. Brother S placed his fish in the box and wagged his head wryly at the world.

You will find it difficult as a moocher to make any permanent peace with the wind. "The only thing good about it," Brother S will tell you, "is the relief when it lets up. Rain – that's different. Staying dry in good rainclothes is fun, like shrinking down in your sack listening to the stuff rattle the tent roof. But wind – you can't keep the boat straight, the lines run under, you can't feel what you're doing. There's all kinds of winds and they're all the same – all bad."

We do acknowledge one exception: a wind that makes special music. I admitted to it in newsprint this way:

> That huge king ran 100 yards horizontally, stringing my 20-pound line taut above the choppy surface, and on that fiddlestring of nylon the breeze played a clear and melodious tune. What a fish! Brother, that is real music.

From whence come the varying pitches in these songs? From the differing lengths of fiddlestring between the rod guides, I've begun to think. But you may never find out. The fish will keep you from concentrating on the melody.

Wet, Windy Days

We lounged in tranquillity in our tent, coffee mugs in our palms and music in our ears: the wind singing through the rain outside and the gasoline stove wheezing out its warmth within.

"I am going to work," Brother S declared abruptly.

We were using separate boats that year, and so he could carry out this rotten threat independently of me. However, there were a few fish around, and I dutifully dredged up visions of him bringing back a boatload while I grew soft with comfort. When he left our rain-spattered cove, I was close behind.

We didn't go far; five minutes to The Pup Island, a green dot in Clover Passage off the south tip of a larger island, Betton. The water between is Pup Pass. Awaiting us in these environs was a catch destined to be immortalized as The Great Pup Double

The "work" of fishing did qualify as work that day. The downpour stopped, and we optimistically removed our rain gear. But the wind – a nuisance much worse than rain – persisted, and the seas kept us scrambling. After a couple of hours I pulled my gear, aiming to go back and resume growing soft. Brother S is a determined man. He bounced, pitched, and yawed in four-foot seas, rods undulating and two 20-pound lines whipping in and out of the swells. A big wave turned him crosswise. As he gunned the motor to gain steerage, both rods arched over.

He threw the motor out of gear – and was promptly crosswise again, with one line scraping down the bow and the other hung up under water on the motor. He got No. 1 clear of the bow all right – but then what? Stay with that one or park that rod and try unraveling No. 2 from the prop? I thought he would go for the prop. He didn't. He theorized, he told me later, that "one in the hand was better than one wrapped in the machinery, because that one was bound to get away anyway."

In 10 minutes he netted No. 1, planting net, pole, fish, and all in the bow, and then turned to No. 2. He tipped up the motor and reached out to feel the line. It angled sharply away. "He might still be on!" he shouted.

The propeller had to come off, preceded by the cotter pin and the nut. The motor was long-shaft, demanding additional reach. Brother S hung far out over the heaving sea, stabbing with his pliers. The boat lurched toward the foaming rocks of The Pup. The rain returned with a vengeance.

What Brother S did can't be done. Fortune didn't just smile on him – it bestowed its blessings lavishly. He didn't lose anything: not his balance, not the pliers, not the cotter pin, not the nut, not the propeller, and not the fish. Some vagary of wind and current took him off the collision course with the rocks and into the shelter of the pass. He tossed the propeller inside, unwound at least four turns of line from the propshaft, picked up the rod, and, while the spruces waved applause in the sailing rain, enjoyed a perfectly ideal, sopping wet battle with king salmon No. 2.

He landed it with his gaff hook and, as the rain stopped again, plunked it down atop No. 1. They were 32 and 25 pounds.

We rehashed the episode while the rain spasmodically bombed the tent that night. It was the first of quite a few rehashings of The Great Pup Double.

There has been substantial time for rehashing; the year of the Great Pup Double was 1957. The thrill of it stirs still. You'll find, in fact, that although experiences may be repeated, the first time around is the special one.

The Great Pup Double was a further first.

"It was," says S, "my first lesson in slack lines. Time after time since then, slack lines have paid off. When there was no fisherman pulling on that salmon, he didn't pull either. So, when you get two on at once, just give one some slack and sometimes he won't go anywhere. The idea works on kings and is even better on cohos – they're great at making their own slack line anyway, by running for the boat. A tight line is all right for tiring out a fish, but not for keeping him on the hook."

THE SECOND ONE: He landed it with his gaff hook

And Brother S has stuck for all time to his theory, formulated in the wind and rain that day of The Great Pup Double.

Cold Days

You've caught fish when the line was freezing in the guides, the sleet quickly freezing on any available still water, and your fingers freezing to each other – caught winter steelhead, that is. Summer salmon should offer less discomfiture. But don't count on it....

Except for one problem, your setup is perfect. The cohos are ravenous. And they are the big ones of mid-September: 10 and 12-pounders. Quite a few hit 15 and one that just leaped looked like 20.

The problem is – the ice-cold rain streaming through your fingers has turned them painfully useless. Working against your chest with the heels of your hands, you mash a spinner onto the hooks. This messy bait is good enough to please these willing cohos, but when you hook one you can't turn the reel handle more than a half-turn at a time. From the hot bite you get two fish. You should have had six.

It doesn't have to be September, either. In a chilling rain in mid-June, after finishing a skirmish with some kings in a favorite hidden pocket, we pulled around a bluff and ran into Clover Pass Resort for gasoline.

Two other shuddering fishermen in a rented kicker boat were arriving at the same time. They were Californians. They had camped through British Columbia and were spending two weeks sport-fishing out of the resort. One, an engineer or scientist of some sort, was doing some analyzing.

"There is thick cold and thin cold, and this rain is thick cold," he said, painfully shaking his shriveled, dripping fingers. "Take that creek we camped at on the way up. In that thick running water the beer got cold ten times as fast as in a refrigerator. Water is thick, air thin. Tuesday morning was cold, but clear. Thin cold. My hands lasted an hour. Today in this thick cold they froze in five minutes." (Confirmation: According to a Washington State boating safety brochure, a human body loses heat 23 times faster in water than in air.)

Thick cold and thin. "That man – he knew what he was talking about," said S. "He'd learned about hands, too. Hands – you have to keep them warm so they'll work, but to work you have to have them out in the cold...."

And there is, you will sometime discover, a cold thicker than rainwater cold: summer snow cold – cold that clings to your fingers instead of running off. In northern Southeast you can expect a sample on any given year.

I endured one on a late June afternoon off Point Retreat. The kings were biting heartily and I had two 25-pounders in the boat and another on the line when came the slush-storm. Here was cold not only thick but in constant replenishment. The action grew progressively slower, my 10 thumbs progressively stiffer, the cold progressively thicker. At netting time a customized technique was required, consisting of an armpit clamped on the net handle and forearms hoisting the hoop to dump the prize inside.

There was now an inch of slush in the boat, and two inches on either side of the fish, splattered there when I slopped him down.

In southern Southeast you may or may not ever have such an experience. The temperature averages but a few degrees lower than down south on Puget Sound. You can always dress in layers enough to keep warm. Except for your hands, that is.

"Mittens and gloves – they're no good," says S. "Like washing your feet with your socks on. Get all wet in the rain anyway."

We tried warming our hands in the outboard motor exhaust water. They got horribly chapped. Rubber gloves in the exhaust water? Hey! That worked. Has to be one of the best ideas we've had yet.

S argues, however, that fights with fish deserve bare hands despite the weather. "Once," he confided to a newcomer who was getting some tips from him, "we thought we'd have to somehow stop burning our thumbs on our reels. We weren't using star drags then – more feel that way.... To save my thumbs I fixed up this leather pad that swung against the spool. It was awful – separated me from my work. I ripped it off and went back to burning my thumbs."...

Thick cold, thin cold. The darling of thin cold must be the cold of a clear June dawn....

From our tent in Knudson Cove we brushed through the cold-stiffened huckleberries and across the clearing to the floats, our breaths fogging white in the half-light. We stepped gingerly on icy planks, and they broke the stillness with loud croaks of protest. Climbing into our 16-footers we were already well chilled. "And we still have to handle that frozen bait," I wheezed. "It didn't used to feel anywhere near this cold down on Puget Sound."

"It wasn't 2:45 in the morning there, either," he said. "Just wait till the sun comes up."

By the time the sun did come up I had exercised on two kings and was reasonably warmed, with a sunny day ahead and with bolstered appreciation of the Californian's two kinds of cold.

Hot Days

Hot, bright days, some "experts" say, ruin the fishing: the light hurts a salmon's upward-looking eyes and drives him deep.

"Don't listen," says S. "Salmon bite when they bite, not just when somebody thinks they should. The ones that surprise you in bright weather are the ones you really remember. And besides, you get to float around out there with your shirt open."

One time I tormented Bremerton *Sun* readers thus:

> The sunshine is in its fifth beautiful day. Our two little boats plane happily over flat blue water, going fishing anyplace. The wind is warm in our faces, something a fisherman feels but seldom, even down there on Puget Sound. Incongruously, the white of winter still lies thick on rock slopes close above. How the air can stay so warm so close to snow is a mystery.
>
> The visiting sport fishermen in the close-to-civilization vicinity of Clover Passage are numerous these days. They come from far places. You can guess which place. The Southerners have unbuttoned their mackinaws and the Midwesterners are in shirtsleeves. The Alaskans have stripped to the waist.
>
> In the few spots where rocks and boulders give way to beaches, swimmers vie for space. Off the launching ramp at Knudson Cove, youngsters yell and splash and are imperiled by boatmen trying to use the ramp for its intended purpose.
>
> The creeks have become trickles. Ten-foot culverts are barely wet at the bottom. Folks with roof-fed cisterns fret and worry they'll go dry. The flies buzz busily down on any aromatic scrap of exposed fish flesh. A citizen tells the local radio station his thermometer is "standing at 85 and it feels hotter." The editor of the *Ketchikan Daily News* checks the wire to see by how much he can say Ketchikan's temperature beat Seattle's today.

Of all the heat waves over the years, the most memorable for us involved an Oregonian named George – George and some "new fish," the term for an arriving run.

George was visiting his brother Ken, a Ketchikan businessman whom we knew well. Ken asked us to "give George a treat – fishing with a professional." Ken had done favors for us and we were pleased by the chance to reciprocate, not to mention by the tenor of his request. Brother S and I were in separate boats that year, so we drew lots to see which of us would serve as the professional. I "won."

Our brilliant skies supplemented conditions already rotten. Salmon were non-existent. Sports fishermen and a couple of charter boats were consoling themselves with red snapper (yellow-eye rockfish) and other succulent bottom fare. None of the tricks I knew for taking the odd fish would take absent ones.

George proved an able hand with mooching tackle and a fully reasonable sort otherwise, and it became evident that fault for failure would not be laid mercilessly upon me. By 6 a.m. the pressure was off and we had fishermen's banter going.

"Sure is easy to sit here," George said, leaning back in sunglasses to watch an eagle soar across the blue and settle atop a spruce. "Doesn't look like any fish will bother us, either."

"I can almost guarantee it," I said.

George sunk his spinner 200 feet to the rocky bottom. "Down there," he said, "is where any fish is supposed to be on a day like this."

"So you've been briefed on the rules," I said. "Not all the fish have, though."

"By gosh, this one has!" he said, and set the hook in what proved to be a 29-pound chinook: fish number one.

I was working the top with a half-ounce of lead, testing for fish defiant of the rules, those that come to the surface on the brightest of days. (For more about these wild fish, which Brother S and I had learned to try for frequently, see "Maverick!" on page 127.) My spinner was in plain sight 30 feet behind the boat.

"Hey!" said George. "There's a fish chasing that thing! Hey! He's got it!" My reel whistled. Fish number two.

George got no more action deep. "Try the surface with me," I said. Halfway up he stopped and yanked. Fish number three.

The middle water gave up another chinook and the surface

layer still another. In the afternoon George flew back to Oregon with two iced chinooks dressing 50 pounds, and I still had three in the box. In the heat of the evening I landed a final one. I rehearsed the glowing account I would give Brother S of my professional capture of six fish, but he spoiled it. He had five himself.

Foggy Days

On a socked-in September day in 1981 near Point Higgins north of Ketchikan we were trolling along cautiously, keeping land in sight. Against the beach we spotted a young fellow in a 14-foot runabout. We pulled in. Was he okay?

"Fine, "he said. "About to run over to Vallenar Point. There's fish there."

"You can't see anything."

"It's only two miles. Easy compass course."

Brother S, who has been known to make reference to "those young days of my own, when I had ten times more life expectancy and one tenth the regard for it that I have now," reminded him that the two miles crossed a main shipping lane, Tongass Narrows, "and you can't hear what's coming when your motor is going." But youth would be served, and away he went, and thousands to one he made it safely. As his admonishers, though, we were mindful of words of caution we had once received from another fisherman. He had headed across this same lane in a soup and:

"Ran between a tug and barge. Couldn't see either one. Ran into the tow line. Didn't know for awhile what was going on. Then the bow of the barge showed up, 10 feet high and hanging over, fixing to roll me under and grind me up. Barely got back out of there in time."

Some of your most eerie mornings will be those you spend shouldering through Alaska fog. Brother S says fish are inconsiderate "because they still bite in this stuff – make us come out and work in it."

You won't likely ever get cordial with it. Our own intolerance began with interruptions to our fishing in the ferry lanes between Seattle and Bremerton in the 1920s, and improved none in Southeast. The problem isn't really that you feel in danger of being run down by a ship. In the non-navigable water where we like to fish, the risk of that seems zero. Break up on a rock? Not likely, at mooching speed, in the calm that often accompanies the soup. Get lost? No, we don't stray far enough for that.

But your spirit suffers, confined in the moist gloominess. You grope around in the hanging, off-white fluff, and you face another kind of mis-orientation when it lifts. Except for certain variations within this process, the scenario repeats itself season after season.

Of the variations, the most notable one blessed us when two big chinooks, close to 40 pounds apiece, gobbled our spinners at once. We pursued them in unknown directions and at length – at one point S remarked "I wonder where we are going?" – and landed both.

So where had we gone? We had no idea. There was no clue anywhere. And when the stuff thinned and the world began returning, we went through the strange confusion of the unveiling. The emerging rocks and trees of the shoreline had to be the same ones we looked upon countless times, but we had never seen them before.

Aspect is the problem. When you are suddenly unblindfolded, your liberated senses balk at the unknown viewpoint. Eventually something prominent and familiar like a certain eagle's nest in a tree stands out and – oh, yes! now I get it – everything gets realigned. You have a fix again.

So presently we had one and, with our two hefty chinooks, were back in the known world.

Fog begets traffic problems not involving ships, barges, or relocated rocks. The number of places fisherman can readily find decreases, and the result may be uncommon concentrations of a half-dozen or more boats, all invisible to the others. I was fast to a good chinook and he was slowly towing me through dense nothingness when an extra force was added to the pull. What was going on? Well, this was: A wire line from a handtroller's gurdy had crossed my nylon and was hauling it forward....

"Look out!" said the fisherman in the boat now linked to mine. "No, you look out!" I said. At this stage we were close enough together to see each other.

What had happened and why became clear enough upon review in an atmosphere tempered by a fish apiece. Hand-gurdy fishermen, like power trollers, keep right on going when a fish lays on, and this hand-gurdy man had picked me up on his port pole while preoccupied cranking a fish on his starboard gurdy. We parted without deciding who should be mad at whom....

Among the nasty features of fog is its tendency to thicken in September, when the silvers (cohos) are attaining their maximum numbers, size, and ferocity: irascible 12- to 20-plus pounders.

"You can't trust those cohos anytime, and fog brings out the worst in them," S says. "Some are violent and some sneaky."

Fog, he says, either increases sneakiness "or your foggy mental state makes it seem so." One of your lines quietly comes abeam in a loose sag. You reel in the slack – and the coho that carried it there turns appropriately violent.

Some silvers employ semi-sneakiness, and they may combine this with another favorite tactic, leaping high in the air at first hookprick. Your reel sounds no alarm. What these fish do is softly bite the bait, feel the hook, and clear the roof. You see no flash but hear the splash. Hey! A fish just jumped! Then the reel does scream. So that's why he jumped!

We hit fog and sneaky outsized cohos together off Grindall Point in September 1979. We were working compass-and-depth-sounder courses. We couldn't see anything. But Jehosephat, what we could hear: great splashes somewhere in the murk astern, followed by screaming reels in the boat. We got 15 of those cohos in one hot hour, several of them above 12 pounds dressed. Two of them, weighed together, totaled 34. That was one fog we forgave.

Calm Days

A few times a season you will be privileged to bask in that gift of nature, neutral weather: not raining, not blowing, not hot, not cold. Totally comfortable. Priceless. In ease you float along on flat water while the eagles soar and the hemlocks smile. Boat and motor respond to your any whim. Barely moving, you keep perfect steerage and control. Your depth-sounder readings are firm and precise, your touch delicate. You feel-fish any depth.

Mediocre fishing now peaks out. A hunch tells you to brush the side of a jutting boulder and you safely brush it. Good hunch. A big chinook lays on.

Now this customer gives you a new batch of delights. Stand up in the boat in balance undisturbed, step from gunwale to gunwale. There's no bracing of knees, no tumbling against the bulkhead, no barking of shins on seat edges. In the security of a stable stance you reach out and net the glistening battler, dump him on the floorboard, and kneel alongside to flip out the hooks, simple as on a kitchen linoleum....

Like other gifts of Providence, these calm days may present themselves unheralded. They may come on the heels of a big blow, or before one, or between two. Do not ask why you have been favored. Just go fishing. Strip the line through your fingers for the sheer joy of it. With the confidence these days bring, you can entice the most indifferent of fish.

The time Brother S and I best remember proving the point was in a July calm in 1982. We were on a Friday-through-Sunday "weekender" along "Fisherman's Row," the east shore of Prince of Wales Island. For 10 solid hours, in recognized big king water off Island Point, we worked in delight and diligence. The tide was in neap stage and the currents minimal. We put our boat and our fishing lines exactly where we wanted them.

We raised nothing. At dusk we rounded a bend into that safe and quiet retreat of fishermen, Trollers Cove, and in quiet contemplation ate and turned in....

"It's six o'clock!" Brother S exclaimed. "The sun has been up forever. What are we doing here?"

The conifers around us remained motionless to their lofty tips. "We must've done something wrong yesterday," said S. "There's no excuse for not catching fish in water like we've got now."

By 6:30 we were back at the point. The calm remained. In 50 feet of water or 150, we bumped our gear along bottom, keeping the motor barely idling or even turned off. Silence. No bait motion except whatever we imparted with the rods. At 100 feet, 2-ounce sinkers yielded clear communication with the bottom.

A nudge sneaked up the nylon and ticked into my palms. It was almost imperceptible. In this unrippled water it was unmistakable. It came again, this inspection-style mouthing by a chinook salmon, and I hooked him solidly. Forty-one pounds. By himself he paid all the expenses of the three-day trip.

Before noon, from far beneath that still surface, we coaxed forth three more nibbling soakers. Then an afternoon breeze stirred the world and took the spell away. Only a ripple – but the ham and eggs would taste good in Trollers Cove.

The break was longer than most, product of the letdown in chemistry that has to come after sessions like these. The trouble with perfect days, Brother S will tell you, is that "you don't like to put up with normal working conditions again."

All of the Above

Mid-morning. Late June. Thick gray clouds were glowering around, but the sun grinned at us from a blue patch, and we doffed our oilskins and blazers. That was the signal for a 25-pound king to hit. The sun disappeared, and three times as many BTUs went with it as I could generate on one lively fish. Brother S, meanwhile, got his outer layers back on. He hooked another salmon and, under thick wool and waterproofing and a resurgent sun, perspired lustily.

Hot days, cold days, wet, dry, windy, calm, clear, and cloudy days – all compete, seldom willing to let any of the others long prevail unchallenged. "Never wear one-piece longjohns," Brother S will warn you. "You have to be able to get the top half off, or suffocate." It is not only, he adds, that the changes can come suddenly. "Something about this country makes you notice them more. Fifteen degrees feels like 50."

These classic hot-and-cold days present themselves frequently, so over the years many of them are bound to include fish or other excitement, or both. And in memorable combinations....

All afternoon the clouds and the sun and the rains and the breezes came and went. The salmon bit. Sometimes shivering when we got them on and sweltering when we got them in and sometimes vice versa, we fed the fishbox. Chinooks (kings), cohos (silvers), and pinks (humpbacks), all were coming aboard.

Of the day's incidental events, one involved a horsefly – or at least involved one of the ¾-inch-long critters we've always called horseflies. Whatever. For familiarity with these horseflies you need only to so prosper as an Alaska moocher as to graduate from an open boat to one with a half-cabin. From nowhere, dozens of these horseflies will materialize inside your windshield.

Usually they treat us civilly, as all but one did this time. We slapped at the stragglers buzzing the cockpit and they merely dodged. Never before had one retaliated.

Brother S had just netted a coho and was removing the hook.

"Ouch!" he yelped. That outlaw horsefly raised a welt like a yellowjacket's on his wrist. The rest of the story in retrospect is comical. Brother S not only rubbed his lump between fish the rest of the afternoon but, whenever action took him the two steps across the cockpit, he limped. "My arm hurts," he explained.

Mid-afternoon produced a fog, or at least an overcast, the bottom of which drew a straight horizontal line 50 feet above the water, and in the slice of clear space between, a seaplane whizzed by. The islands around us were no longer mountainous. Beneath the soup they looked instead like flat grass prairies. Then the curtain wcnt back up and thc sun madc a rainbow in our spray as we planed to a spot we like at dusk, and stopped there.

We had our daily fishing contest going (do not suppose that since you and your partner are splitting the take you pay no attention to who catches more or bigger fish). We rigged bait and maneuvered gear at our best, but neither of us could get much ahead at any time. And at the charmed final hour, as happens once in awhile, we were dead even. We had three chinooks and five cohos apiece. The chinooks ranged from 24 to 34 pounds.

"Think we'll get one more?" I asked.

"Think so. Bigger'n these, too."

"Who'll catch him?"

"Probably me. Wup! He's nibbling right now. Wup! He's on!"

The sun was in a blue patch, just above a distant tree line, but still warm. Ten minutes earlier we had thrown our coats and rain gear into the cabin for the tenth time. Now huge drops of rain splattered on the deck. "Here it comes!" I said. "I'll get our coats – n-no, guess I won't. I've got a fish, too."...And so, back to back in a buckshot rain, we cranked in those two salmon.

The rain stopped. We dripped. Two big, wet fish sparkled on the deck. Brother S hoisted one on the scale-hook and I read the numbers: "Thirty-eight pounds." Then the other: "Thirty-nine pounds. I knew mine would be bigger."

"Yours? This one's yours."

"No, this one's mine."

"No, yes, no – you mean – we don't know which is which?"

We didn't, but it wasn't our fault. The mixed-up day did it.

3. DYNAMICS BELOW

We were refueling in Ketchikan when another boatman pulled in, disembarked, and resolutely approached us. He had something to say and we were the available listeners.

He was a retired General Motors supervisor from Grand Rapids, Michigan. He had towed his 23-foot inboard-outboard cruiser behind a motor home to Prince Rupert, British Columbia, launched it there, and come the 90 miles north through fog, rain, sunshine, and wind. Yet what he wanted to unload about was not the weather.

"This crazy business of tides!" he said. "This crazy salt water! While I was on the launching ramp, it came up and flooded my trailer hubs. When I anchored, it went back down and almost left me on the rocks. And on the way here it pulled me along part of the time and pushed me backwards the rest. It even looks different one place and the next. It's crazy!"

If you were born to crazy salt water and never have been without it, you will share our surprise at his surprise. Tides – what would we do without them?

High Water

When's the fishing better – during big spring tides or lesser neap tides?

If you say the neaps, we say you're right. But be ready for the exceptions. Turn a circumstance. Figure a way, for example, to parry the great accumulations of kelp, seaweed, logs, lumber, and junk which the extreme highs lift from the shores and carry to where you want to fish. Pity the power trollers, who collect the stuff liberally on their lines and lures. Mooching, you can lower your bait through a hole in it.

At Island Point, off the east shore of Prince of Wales Island, where big drifts and big salmon come together, Brother S strained his ingenuity. We cannot say why we so often choose this place for fishing when the junk is at its worst, but so it seems. One big mat covered a quarter acre and was so thick, the top layer had dried in the atmosphere of a few rainless days. This floating island was dense enough to stop a boat.

S surveyed the snarl from close range, cocking his head in contemplation.

"Salmon hide under cover, and one is hiding under that mess now," he said. "Let's go after him – sink our spinners here, circle to the other side, and pull 'em in under."

This plan, as you may deduce upon reflection, had its limitations and enabled us to get "in under" only a short distance from the edge – but it was enough. Brother S connected. He brought in one fighting salmon, 30 pounds, and one load of seaweed, about two pounds, strung on the line by the travels of the hooked fish.

Also on these tides, beware those wayward logs which, heavy on one end, assume vertical posture in the water. The ones that stick up high enough to be easily seen pose little problem – might even be a benefit. Sometimes one will remain in an area for days in defiance of strong currents, and there are fishermen who claim that whatever eddies or forces control them also retain salmon.

But fear the deadlier deadheads – those of practically neutral buoyancy, the upper end washing at surface level, almost invisible. Pray not to hit one at high speed in a mooching-class boat. And be thankful there aren't more of them.

Of other low-floating hazards there are plenty. And in a choppy sea in imperfect light you cannot miss them all.

What sensation rivals the one you get at 20 knots from the sudden *thump*! against the quarter-inch of fiberglass separating you from the cold, deep sea?

The thump is the first of two; between them is a rattle. The first thump probably does no damage; the bow goes up and over.

The rattle is the whatever-it-is tumbling along the keel, probably doing no serious damage there. What you fear is the second thump, the one at the propeller, due in another fifth-second, just long enough for you to close your throat against your heart, yank back the throttle, and hunch up for the worst.

The thump comes. The outboard leaps up, the boat wallows to a halt, and the cause comes rolling up in the wash. Relief. It is only a 2x4 stud. A few splinters bob around. Inspection. Prop still O.K. Skeg slightly bent. Carry on....

If schedules oblige, avoid the flotsam by following a ship, because ships clear and calm the water in their wake – though sometimes not completely. Vigilance always!

Other hazards include plastic bags and wrappers, illegally discarded on the water. They drape themselves around your water intake, and the motor overheats and quits. Will it start again when it cools off? Yours does, and you breathe your thanks. You also reaffirm your vow never to throw plastic overboard.

...Less stressfully, check out the evidence of big tides past: storm-blown logs, roots intact, deposited ashore so long ago they are now brush-grown. We have two favorites among these, Nos. 1 and 2 we call them.

No. 1 is a landmark over a lively stretch of fishing water: a whole tree, three feet in diameter, stranded on a rocky point, roots to the waterward and bole swung inland. In our 40-years-plus it has not moved.

What a story it must hold of its arrival at this high resting place, 50 or 200 years ago. That big tide must have enlisted the aid of a whole gale, to boost our landmark three extra feet with blown-in Pacific water and three more with surf (or possibly a tsunami?).

This log testifies not only to the height of water but to the durability of Alaska cedar. It is still sound. And it has one more grand feature. One of its bleached roots bends up and over to point like a curved finger at a deep spot close to shore where we have hooked at least a dozen big chinooks.

Log No. 2 is special because it marks the margin of a rare stretch of sandy beach where we easily can, and every year do, go ashore to admire it. How long ago did it get stranded? Perhaps at the same time as No. 1, and at least so long ago that, when we first noticed it more than 40 years back, it was already the backdrop for a patch of lupine flourishing on its seaward side.

SO HIGH A TIDE SO LONG AGO: Here's the scene described above.

As to high tides otherwise, count your itinerary incomplete until you have poked around or into a flooded stream mouth. The fishing (if legal near the particular stream – check the regulations) just might be spectacular. If it isn't, reap the other rewards.

We began doing this in early years, at a creek-mouth in Bond Bay on Cleveland Peninsula, and haven't stopped since.

Beneath the limbs of alders where salt water seldom came, a school of needlefish was congregated. "Feeding on bugs the high water washed out of the bushes, I bet," said S. Perhaps. Outside the needlefish were numerous dogfish, pack hunters known to drive their prey to dead-ends. (When dogfish were valuable for their livers in the 1940s, one fisherman told of making $2000 on a pack stranded with a school of herring in a tidal pool.)

Knowing that dogfish and salmon may feed together, we

fished off this stream mouth. We caught dogfish. But later, further out, we also caught salmon: salmon full of needlefish.

In the meantime, in our two 16-foot boats, we poled up that stream, around bends, beneath overhanging trees, and between narrowing grassy banks until the only way out was backwards. Under a canopy of leaves in a silent forest, we ate ham sandwiches and drank from the creek before going back to work.

Another "navigable stream" we've ascended – this one together in a 20-footer – is one that feeds into Trollers Cove, a famed fishermen's haven hooked into Prince of Wales Island off Clarence Strait. If you've never cruised the bayous of Louisiana or the Okefenokee Swamp – now you don't need to. Probe into the Trollers Cove woods on a bumper tide instead.

We never got lost – although that is a possibility, the water being deep enough for travel even off the main creek channel. We always knew which way was out. Good thing. When the water started to fall, Brother S was suddenly all business. "Right now," he said, "we're about 18 feet up. And right now we've gotta get out of here – or who knows when we will?"

So – have fun. But don't misread your tidebook.

Low Water

If you want a foolproof formula for big salmon catches on low tides, we can't help you much.

We've had feasts and famines. The feasts are worth the wait.

Brother S, a student of shallow water fishing anyway, theorizes that receding water should mean "thicker fish in what's left." He does admit that "theories aren't fish."

But fret not when no salmon offer.

The Panhandle's uncrowded attractions include its tideflats, vast and inviting: acres of privacy. Stow your gear and go loitering where the crabs scurry and the clams squirt. Crabs? We've landed plenty for dinner with an oar.

Cash in all around. Cherish especially those rare days when big runouts coincide with sunshine beaming down over the evergreens, intent on noble purpose.

The soupy sand warms to extended exposure, and thick vapors rise to tweak your nostrils.

Do not let this opportunity escape! Doff your boots and socks and be a child again. Stop, stand, and jiggle. The ooze between your toes doesn't differ much from what you've known elsewhere, but it feels better here.

If you get around to fishing, be ready for anything.

For our first several years it seemed to us that the salmon vanished with the retreating Pacific. Then, without warning, in 1964, came the first of the feasts.

The tide had fallen to near minus 4. In five feet of water we were peacefully trolling, unmolested by fish, in a smooth-bottomed part of Vallenar Bay. Crabs scooted beneath us, but we had loaded up on them the day before.

Vallenar Bay faces north. The flat bottom at the east turns at the west into a conglomeration of kelp patches, reefs, and rocks, small, large, and gigantic. We had left the flat and were skirting this foul ground, approaching South Vallenar Point, when a huge fish jumped between a couple of rocks to shoreward.

"Let's go in there," said S, the shallow-water man.

"Too many rocks," I said.

"No problem," he said. "When the tide's down like this you can see them all."

We have never settled the question therein presented. We have just quit arguing it. Aren't we dropping down within range of "new" rocks? But this big fish had jumped, and so, cutting down to one short line apiece, we went in – and had no problems, except with fish.

That foul water was infested with king salmon. The first two hit at once and went in opposite directions. And in a place like that, where they can't go down, they go a long way out. We couldn't follow either one. Both reels were shrieking. Brother S's fish slowed down first, and he turned to ask where I stood with mine, timing this query just as my fish jumped. "Holy smoke," he said. "He's out on the horizon. "You got any line left? Wup! I gotta work on mine."

When we got his fish and then mine near the boat, each was being followed by another. Both of these departed when they saw us, but the hooked pair came safely aboard.

The bite lasted only the half-hour or so while the tide was slack, but in that time two kings in addition to the first two accepted our invitation to the fishbox. The rest had to stay frustrated because, with mooching gear, we couldn't accommodate them fast enough.

STRETCHED NYLON: How far is my chinook going? "He's out on the horizon," said Brother S. "You got any line left?"

What do you do to tap the get-rich-quick potential of a setup like this? "We'll get some 100-pound outfits and just throw these 30-pounders aboard," says S. "Now we spend 20 minutes on one fish while the rest swim around with their mouths open."

The 100-pound gear has never materialized.

Another of our low-tide bonanzas was with late-June cohos, five pounders to which our 20-pound gear equates like our envisioned 100-pound stuff to chinooks. These June silvers were also in shallow water, less characteristic of them than of kings. They were so thick we could see them swimming around the boat, whether or not we had a "decoy" on the line attracting them. We got 40 in an hour, a – rare hour even in an Alaska coho run.

From these and other low-water hot periods, no pattern has emerged. When you come up with one, let us know.

Fast Water

The boat swapped ends in the rip and Brother S swung his line over my head for clearance.

"We've been getting into this stuff every year for 30 years," he muttered... "Forget it in the meantime and then get surprised all over again."

We were running a salt-water rapids, a millrace unleashed by a jumbo spring tide in Southeast. To test your mooching ingenuity, figure out how to operate when these days of rushing water arrive together with a run of big chinooks. The theoretic peak of the natural forces responsible occurs on June 21 or 22, but anywhere close can inconvenience you sufficiently.

Expertise gained outside Southeast – on Puget Sound, say – won't likely prepare you. In Seattle the tidal range is about 17 feet. In Ketchikan it exceeds 23 feet, with highs above 18 and lows below minus 4. Be thankful for small favors. Southeast tides are modest compared to some in Alaska. Other good news includes the possibility of great catches, if you guess right.

More bad news first. Don't be confident about making much progress as a monster-tide fisherman this century. Time is short, and we haven't made great progress in 40 years.

"The trouble is that we don't get enough experience," says S. "Tides like these only come along a couple of times a season."

In reviewing our logbooks (a fond winter pastime) we note assorted complaints about big tides. One of mine calls the water "wicked." S gets more specific – "Couldn't feel the bottom at 50 feet with five ounces. Sure couldn't tell when anything bit."

The difference between a 15-foot run and a 22-footer, as S describes it, "sure isn't just the difference between 15 and 22. At maybe 20 feet, everything changes."

Indeed it does. Normal flow routes get overtaxed. Thresholds are exceeded and cockeyed new patterns created. Powerful eddies swirl where normally there are none. Violent, maverick tiderips crop up. Whirlpools sizzle. In the swirling depths, counter-currents wrap your line in birdnests.

For one example of Jekyll-Hyde transformation, take the trauma we got into at normally peaceful Lumpy Point.

Lumpy Point, so named by us though undesignated on charts, demands a moocher's love at first sight, and demanded ours on first visit, in 1957. Beautiful. A classic example of the realm, a rough arm of rock thrust out into deep water, with higher knobs topped by tufts of grass and a few stunted trees, weather-toughened versions of the rain-fed specimens mantling the hills.

The current was at a fast troll across the point and a slow mooch along the flanks – the perfect place to sink a spinner to any lurking salmon.

It proved up on its invitation, yielding three chinooks that very first occasion. There would be many more in years to come.

The salmon, per a not uncommon trait of theirs – changing hangouts with the tide – exercised the prerogative here. They liked one side on flood, the other on ebb; always the lee. So they were there all day. Ideal! "Places like this," says S, "are what we came for." But not during monster tides...

In the maelstrom along the normally peaceful flank we could not get our gear down until we piled on six ounces of lead – and then couldn't tell whether anything was nibbling on it.

What to do about times like these? We have a couple of acquaintances who corrupt themselves by rigging up with big rotating flashers and trolling fast enough to stay strung out. Brother S refers to this kind of fishing as "whirling" and the fishermen as "whirlers," and views the operation with all the zest of a dry-fly purist encountering a worm man on an elitist eastern trout stream. "But," he concedes, "some of them do catch more than others, so I guess they have a skill of some sort. But they can have it."

Other fishermen who "can have it" are the downrigger types.

With one aspect of downrigging – playing the salmon on a thin, sinkerless line after the fish snaps it free from a 10-pound downrigger ball – he has no quarrel. "But you have no idea what's going on until all of a sudden the pole is jumping. That's not fishing." Brother S filed no protest when Alaska outlawed downrigger-sport rod combinations for handtrollers.

What else can you do about excessive tides? You can experiment. In fact, you have to, since standard procedures don't work. After enough tries, as we demonstrated at least as early as 1964, something will....

...Off a point normally good for mooching, the current mounted to about five knots, and a tiderip formed where never we had seen one before. It was clearly marked by converging clusters of seaweed and junk. It grabbed our bow and heaved us broadside. "No use to fight this stuff any more," said Brother S. "Let's keep going where we're pointed."

The heading led us into an inlet where in the quiet of normal tides we had never caught anything – and where we now hung two big slabs.

That bight was overstocked with chinooks. Hungry ones. Brother S said they had been "too busy swimming to eat."

Brother S' fish shook the hook but I got mine alongside. "You net him," I said. "O. K." he said – "Wup! No, can't – Got another one. You'll have to." I did, and before his was aboard I – "Wup!" – hooked yet another. By the time the tiderip subsided we had four chinooks. Three of them were mine – three perfectly matched beauties, 44 powerful pounds apiece.

Among other big-tide fishing frenzies, the sides of a submerged rock near Caamano Point netted us five chinooks in an hour and the borders of a kelp bed in Vallenar Bay turned out 24 five-pound cohos in 60 minutes. But the monstrous tides are too few. As Brother S says, we never get enough experience.

But, he adds, "It does seem clear that you have to go to areas where you usually don't. Tides that make things different for us make them different for the fish, too. They collect in new places. Find those places and fill up."

Slack Water

As Alaska winters glorify its springs, so the hard ebbs and flows of its tides enhance the soft interludes between. The faster the flow and the more turbulent your workplace, the greater your relief and keen your feel while the sea is pausing to gird muscle for the return trip. "If a king ever wants to bite, this is when," says S.

He speaks a truth you will learn early in your career. Puget Sound taught it to us more than 60 years ago, and Alaska reinforced it repeatedly. One example of Brother S' faith I recorded in a column this way:

> We hit the water at 5 a.m., apprehensively. The kings were in. But a tide outrushing from 17 feet at 3 a.m. to minus 3 at mid-morning was sure to be disagreeable. It was. At mooching speed, swirls ruined our touch and snarled our lines. To try to find bottom was to risk finding but not feeling it – birdnests.
>
> But the kings were in. "They'll hit at low slack," said S.
>
> The magic time grew nigh. Current speed waned. Our senses sharpened. The kings were in.
>
> ...We crept along in still, deep water close to the beach. On the surface ahead lay a fallen hemlock, roots still attached ashore. Off its tip, herring were flipping.
>
> Brother S checked his wristwatch. "Almost low slack," he said. "About time for a king to hit... Wup!"
>
> And *Zinnnggg!*...right on time....

The salmon I was referring to that time checked in at 31

pounds. And before the water got up steam again around the tip of the floating hemlock, we had added two more....

You may never match Brother S' timing, but you will come close. You will catch salmon, and especially big chinook salmon (cohos and pinks are less selective), when the tide slacks.

In June 1978, during three days when we caught a total of six kings, three of them hit within 10 minutes of the morning low slack:

Day 1, June 21: low slack, minus 4.2 feet, 8:48 a.m. Twenty-six pounds.

Day 2, June 22: low slack, minus 4.3 feet, 9:34 a.m. Thirty-two pounds.

Day 3, June 23: low slack, minus 3.9 feet, 10:21 a.m. Twenty-two pounds.

Times like these activate the slack-tide *sensation* in you, this sensation being something that lies latent in your ribs until aroused by any slack, or even by a funny eddy. You drift into a calm sidepocket and you feel the sensation.

Brother S says the reasons you can schedule the action (and the sensation) for slack tide are these:

1. A fish burning calories against racing water "ought to be ready to eat when it lets up," so your odds improve.

2. Your senses are honed by the sensation. You make your bait work at its enticing best. Your anatomy is tuned to the slightest nibble.

3. You take certain privileges with the data. Modest delays between prediction and strike are disregarded. And "not many people check their watches after the fish bites."

4. The passage of years provides hundreds of slack tides from which to select those to remember.

5. A mystery factor contributes. "No way," says S, "does the water suddenly go quiet everywhere right at tide-table time. Yet over and over that's when the action starts. The fish must read the book."

Do not expect to come up with any sure explanation, but do get in on the fun. When the tide hits the big spring stage and the racing current yanks fiendishly on your gear and ruins your mooching touch, don't fight it. Tread water until the slack.

Deep Water

When a certain rock and a dead cedar beyond it moved into right alignment, I raised my thumb and let the three-ounce sinker tow the spinner down. I had been hitting bottom at 90 feet but here it dropped to 130. "We're there," I said to Brother S.

I hoisted and dropped the tip, reeled the slack, and waited....

Through the thin filament came a cushioned touch. *Tick, tick....* I heaved the tip up, snapped it back down, spun the slack in, heaved again. Contact! The rubbery length of nylon between a 30-pound chinook and a chinook fisherman came taut....

After due process I got the fish up, and S netted him. "Nice fish, even if he was down too deep," he said.

"Not too deep," I said. "Just right. I just proved it."

"No fish should be down that far," he said.

"Where there's one there's likely another," I said – and made the point with a 28-pounder.

"Nice fish," he said. "But there'll be some big ones on top later on. I'll get even."

If you and your partner always agree on everything, your days on the water may grow dull, unless fish are plentiful. They are not always plentiful.

Back in history, Brother S began trying to convince me, and maybe himself, that we ought to do our mooching near the surface – "the work's a lot easier."

"But to mooch you have to go deep," I said. "Feed the gear into holes and hiding places on the bottom. It's the system."

"The system needs improvement," he said.

But now, with my two kings in the fishbox, he turned on the deep-water touch of which he is capable. He sunk his spinner into the hole, came up a few feet, stayed a minute, began a fast-slow-fast retrieve, and, 10 feet from the boat, nailed a chinook....

"All that extra work for nothing," he said. "Good fish. He grabbed it right on top."

"He followed it up from the bottom," I said.

Awhile later I went into the hole and hit another chinook.

"He followed it down," said S....

As related elsewhere in these pages, Brother S' "improvement" does indeed produce fish at times. And while he will always do the extra "work" of going deep when necessary, he expounds enthusiastically on the woes involved:

"You are always sinking and hauling gear: work. You hook too many bottomfish. You snag bottom. And if you hit bottom without feeling it, you let out more line and get snarled up. It's a headache. The trouble is that sometimes you catch lots of fish."

Bonanzas from the bottom – we've had them. But none matches one described to us by the late Lawrence (Kit) Carson, Alaska big-game guide and fisherman, whose home base was a cabin in Knudson Cove, not far from our tent in the huckleberries.

"The kings were in at Grindall," he said. "This was in winter. They were in a hole 200 feet deep and I was putting a hand line down there to get them. One after another I pulled them up – 33 chinooks in one afternoon."

WHAT A HAUL! Thirty-three chinooks in one afternoon, Kit Carson tells Brother S.

We don't know whether we ever found "Carson's Hole" at Grindall. We do work several there, and others at points south along the east shore of Prince of Wales Island. "Fisherman's Row," we call this shore, for a series of pockets (and, to encourage debate, some good top water as well) strung along it.

Fisherman's Row lies across Clarence Strait from the inhabited side, and Southeasters may strand you there for days.

One answer to a problem like this is – don't go. Look for pockets closer to home. They are there. For years they put out for us in and near Clover Passage – practically a suburb of Ketchikan – continuing to do so until the sports fishermen of encroaching civilization won exclusive chinook-season rights to the territory.

"Those deeper places in the bottom – they're virgin country," Brother S admits. "Trollers run right over them – miss everything in them. You have to know they're there."

Some of these deep places are craters; they have no side entrances. "The only way a fish can get into them," says S, "is by going down over the rim. That's the only way a moocher can get in, too. Kings love them. Cohos and pinks not so much. Kings stow away in them, like a man in his den."...

When electronic depth sounders demanded entry into our equipment list, we accepted them with reluctance. They proved to be not all-seeing and therefore not all bad. They do not show all the holes. Their beam can echo from the rims of craters instead of from the bottoms. Good thing. The old hand-feel-style probing of pits and hollows ought never be updated.

Shallow Water

On a day in the 1990s you are still getting acquainted with your wild and diverse realm. You troll around a point, and the rocky bottom beneath you suddenly changes. You shorten line to clear a sandy bottom close beneath the keel.

No use to fish in this thin layer of water, you think.

Ahead, the shape of the green shoreline indicates greater depth. It's only a couple hundred yards. You think you'll just let the gear dangle overboard until you get there.

Zinnnggg! A chinook thwarts your plans. *Where'd he come from? But – no matter now. He's hooked, and.... how the heck far is he going, out there on the surface?*

You'll find out: far enough so you'll chase him with the boat, to save your line and your time – and at last get him aboard....

Shallow water king fishing – here are some of Brother S' comments about it:

"In the first place, it's not the same as low-tide fishing. Something like it, but not the same. Fish can be in shallow water whether the tide is out or not. Some big kings like to come in and check out beaches that were out in the air at low tide. They're like cutthroat trout, scouting territory they can't get to until the water comes back."

And once you've caught a few of these unconventional kings,

he adds, you'll keep checking for them. Still, though you keep getting them 10 or 20 years from now – well into the next century – none will shock or thrill you or live in memory like your first few.

"I hope," Brother S commented to me one evening, "that we're not still sold on those first ones because we're getting senile, like old guys who can tell you what they did as kids and can't remember yesterday."

A better possibility (we hope) is that we treasure first times because we can't repeat them in the same places, due to commercial closures or dwindling runs or both.

The area comprising Vallenar Point, Vallenar Bay, and South Vallenar Point is an example. It also illustrates something else: that our near-shore harvests tend to be enhanced by other memorable circumstances – as if they were not sufficient of themselves!

Case 1

Puffs of breeze began from the east, and we took notice. At Caamano Point, where we were working, east winds rile not just the Clarence Strait side but the Behm Canal flank as well. Bond Bay on the Behm side offers refuge during standard Southeasters, but it is no place for people when the east wind blows.

The puffs became gusts.

"Time to get out of here!" said S. "Vallenar Bay will be sheltered. Let's go!"

We pulled gear, started the big motor, and set about demonstrating that fallacy which holds that small, fast boats enable escape from sudden seas. In the real world the theory collapses before the storm, privy to neither its direction nor its speed of generation. The buildup time in this instance was about three minutes. The Caamano-Vallenar crossing takes 20 – when the water is smooth.

The wind was slightly port of head-on, coming at us at 25 knots or so. The seas kept mounting. We climbed, plunged and splatted, climbed, plunged, and splatted, heaving our wash to the wind. We were down to quarter-throttle, and therefore also sashaying.

We were not worried, exactly; our 20-foot ragtop Sabercraft would hold together. But now we encountered another fallacy –

or at least an unadvertised shortcoming of deep-vee boats. They ride softly? Not when they so sashay that the waves hit the flat side of the vee. Oof! The price of progress was the loosening of innards. The 20-minute crossing took two hours....

Vallenar Bay was a millpond mirroring the trees of Gravina Island. In our new scenic stillness we could have putt-putted along indefinitely, watching the drift of thin blue smoke behind us, and thoroughly enjoyed catching nothing. Catch nothing? Indeed, we often do. But not this time.

The tide was well up. Brother S had a hunch. We made a pass between a kelp bed and the trees of shore, over a barnacled bottom that would be bare at low tide and was in plain sight now. Toward us a grassy point extended.

"Wup! Strike!" said S.... "Didn't get hooked, though."

"Bottom," I said.

"Not bottom. Salmon. Not 15 feet of line out. Light sinker. Bait's gone. Salmon. Let's go back there."

I lifted my gear out of water and turned about. We neared the grassy point and this time S connected – a 30-pounder that charged the grass and then the kelp but finally graced the fishbox. "Same fish as the first one," said S awhile later, turning toward me from the cleaning rack. "Here's the bait he took."

There were more kings under the flat surface of Vallenar Bay. Three of them joined the first one in the next 90 minutes, while, out in Clarence Strait, the sea rolled. With 7-power glasses we could see, or thought we could see, a jagged white line of breakers bombing the black rocks of Caamano, eight miles away....

Case 2

On an incoming tide we headed our 20-footer into a little flat-bottomed bay and, at three-foot depth, dropped the hook into the eelgrass for a break: coffee and fried ham sandwiches, plus the noon news on the radio.

Two fellows in a 16-foot runabout were also here. They finished their refreshments, waved salute-style at us, and departed. We watched them plane out a quarter mile, swing broadside, and troll around a bend.

They missed the fun. We also hoisted anchor but, instead of running out, trolled short lines. The eelgrass was still in sight when the surface swirled behind us and Brother S' reel screamed chinook – a shallow-water individualist who peeled the reel down four times before coming to net.

Case 3

We were in 15 feet of water well offshore when "Wup!" S said, and braced himself with bowed rod and shrinking reel at the gunnel against a 100-yard surface run. I manned the motor in pursuit. "Faster!" said S. "I'm still not gaining."

Then he leaped as if he had seen a sea monster. "Rocks!" he shouted. "Bottom! turn right!"

Every boatman must know the feeling. Conversely, no land-lubber could know it. And you couldn't describe it to him. The good old earth, when it shows up this way, is suddenly foreign, unfriendly, and frightening.

We had little excuse for being shocked. In a topography like Southeast's there will be plenty of places like this. The nautical charts show them. The Coast Pilot says steer clear, unless you have "local knowledge." We had "local knowledge," but obviously in insufficient detail. We also had a big, distracting fish.

Through a rippled surface the boulders wiggled at us eerily. The shore was half a mile away, yet here were these rocks. Were we going to land on one?

But the bottom fell away again, S reeled, and I followed the line at modest speed until we netted a 39-pound chinook.

Case 4

From a tiny gravel beach walled in by rock Brother S pushed off toward the entrance gap and the open water beyond. He hung his line over the stern, ready to start mooching. The boat was about 30 feet off the little beach. The entrance was maybe 20 feet wide. The water was not four feet deep. And that is where a king hit – hit, stormed out the passageway, dove deep, and came up again further offshore, where S landed him....

We can give you no rule why these shallow-water fish do as they do – but they do. Try for them regularly. The first few times you make contact will stay with you from those days to this.

Clear Water

Through five feet of water I could plainly see the brute of a king salmon that was on my line. Beneath the fish, another 12 feet down, the boulders of the bottom were also clearly visible. I could have been in an aquarium. I was in fact in Clarence Strait, off the remote and lonely rocks of the Niblack shore.

Perhaps a few times a season, perhaps but once in several seasons, you'll be treated to such stark visibility below. Clearwater Days, you may call these occasions. Two of the factors that produce them, besides the clear water itself, appear to be a calm surface and an overcast sky, for illumination without reflection. Whatever the cause, what you can see on Clearwater Days will take your breath away.

The tug-of-war between my fish and me was in sustained deadlock. He was motionless and immovable. All the black spots on his back stared at me. With my arched rod tight in my hands I stared back.... *He is four feet long – must be 75 pounds.* And still, what I could see was rivaled by the clarity of the seeing.

From the upper jaw of my fish the shank of my front hook projected. It was solid.

The other hook was embedded in the gill plate from the outside. No way could his teeth make contact with the leader. He was not going to get away.

Clearwater Days do have one drawback. They magnify. The king made several hard runs and then was mine – a prime and beautiful soaker but, at 46 pounds, about 30 short of what he'd have been if I had lost him....

Nature has never announced Clearwater Days in advance, and do not expect her to start. You'll still be suddenly encountering them in the year 2010, not always in total ecstasy. Watching a big king inspect your spinner and coolly turn away will do nothing for your baitcutting ego.

But then another soaker will fall for your phony and you forget the failures.

The salmon you are privileged to get familiar with on Clearwater Days may be cohos as often as lordly chinooks, but spare the tears. Get yourself mixed into a cruising gang of the silvers and you are courting a heart attack. We have faced this risk a few times.

On one occasion five or six cohos, late-season customers of 10 to 16 pounds, were in sight at once. They lurked around the boat but did not offer at the baits. "They're everywhere!" said S. "What – what will this be like if they start biting?

Except for one of them, they never did. But we had fun before that happened. At one point three of them abreast were following one bait, two feet below the surface. One then moved up alongside it, sniffing around. This closer scrutiny ended in rejection; he veered off. Then another approached, but he also turned away.

The third coho drifted back 25 feet, then sped forward. We watched him open his mouth wide, engulf the spinner, do a 180 bend – and set the reel to screaming. He came in at 15 pounds.

On another Clearwater Day we were anchored on High Island Reef, casting. So transparent was the water that we could run an eye down the sloping anchor line all the way to the gray zinc of the anchor, lodged against a corner of dark rock, 12 feet down.

A school of herring moved close and we admired the uniformity of their shapes and sizes and the synchrony of their movements. The black pupils of their eyes made a pattern like birdshot.

As we were absorbing these features, two larger forms materialized nearer the boat: two salmon, quietly flanking the school.

Brother S claims that salmon *herd* herring. "They're like the lions on TV in Africa. Between meals the lions act like they're friends of the zebras. The salmon are friends of the herring, too – guard them and keep track of them until eating time."

One fish weighed about 20 pounds, the other near 30. The big one was in the lead, the other six feet behind. I eased my rod out, drifted my spinner within inches of the big one's nose. It was working perfectly, flashing and beckoning. He did not move.

"Look at that!" said S. "He's crazy! That bait ought to be driving him nuts!"

I worked it back to the second salmon. He, too, ignored it.

From beneath the boat came a flash and a streak – a third chinook, by all odds. He charged out and up, routing the other two and sending the herring skyward. Then – nothing. The herring resumed formation but all three salmon had vanished.

"Strange things like that must go on all the time," said S at bunk time that night. "Today we just got in on the show."

On a Clearwater Day in 1977 we got in on one while trolling

slowly in Grindall Gut, holding our own against the current. I was peering over the side watching the underwater world go by when a shower of stars appeared, 15 feet down. "Herring scales!" I told Brother S, in instant analysis. "A salmon must have knocked them off." On that hunch I cast my spinner aft – and landed a 32-pound king. Again, what dramas go on down there that we never see?

Another show we did see occurred when visibility was exceptional despite the presence of some orange organisms, probably shrimp spawn, in the water. There were herring preying on the shrimp, salmon preying on the herring, and two fishermen aiming to prey on the salmon.

We had been party to this same food-chain story the day before and had caught salmon. High slack tide had been action time, and we were back for another high.

Shrimp, herring, and salmon all were on schedule. We had cut the motor and were casting. We drifted shore-ward. The bottom came up within 10 feet, graphically showing off its gray and greenish boulders and the white barnacles growing on them – and a 30-pound salmon.

You've seen perch twist and turn while picking at barnacles on a piling. This salmon was picking at the barnacles on the rocks!

I lowered my spinner. He came up to meet it. He opened his mouth. The bait disappeared inside. A nibbling sensation came through the line; no hard jerks or pulls, just nibbles. I waited.

Holding or hiding my bait, the salmon turned to stand on his head, waving his tail at me, while keeping his head in one place. I couldn't feel him. It was weird.

Finally I yanked. The hooks came out but the bait stayed in. The salmon swam off nonchalantly. I saw him no more....

We did catch two more kings around the shrimp spawn. We also returned for high slack the third day. But some natural cycle had run its course. We saw no shrimp, no herring, no salmon.

And no underwater picture, either. Rain on the surface spoiled it – if in fact it would have existed otherwise – which I doubt. Clearwater Days are more rare than that.

Vigorous Water

For an hour and a half the action is furious, and when finally it wanes the fishbox is resplendent with three or four glistening chinooks or a dozen or more cohos, or even several of each. You are winded but highly gratified, and a little smug. Not only have you made a week's expenses, but you did it right in the 90-minute stretch of tide when your accumulating experience said you would.

Go through this kind of drill a few times and scientific curiosity may add itself to appreciation. Tides – they're the most important thing a fisherman has anything to do with. Shouldn't you know more about them?

Well, yes and no. Admire tides. Keep track of them. Make the most of them. Capitalize on them. Welcome them. But to avoid frustration, don't try to understand them.

One of the experiences from which I offer this caution I logged in the dull world of salaried employment, in an agency specializing in the testing of underwater ordnance. To correct test-range data for the effect of tidal flows, an oceanographer in my group was tasked to calculate these flows.

Moon and sun affect tide, right? But how many other things do? My colleague told me a sad tale of "all the celestial and terrestrial variables" – of the earth's rotation and its travel among the other planets, of the sizes and shapes of its land masses, of water depths, of the force of Coriolis, and so on... "How many factors in my equation? Not nearly all the possible ones, at this location, but so far, 17."

Seventeen variables! Take heed! If you decide despite my warning to investigate tidal phenomena, vow at least to stick to such simple aspects as why in the world the water rises and falls further in flow-restricting inland waters than in the ocean supplying the push – or why parts of the Arctic coast have virtually no tide at all.

But forgive your fascination. Tides are in the souls of fishermen. Marine and CB radios crackle with the proof.

Complaints predominate :... "Rotten show in this current – be better on the slack." ... "These big tides are going to drive me nuts – seaweed and crap everywhere."... "Never seen fishin' so bad. We need some bigger tides – get the fish moving."... "If there's nothing on the big runouts next week, there's nothing this year."... "This chop is driving me rumdum. It'll flatten out when the tide changes."... (and in fact it sometimes does, although, as Brother S once remarked, "blaming the weather on the tides is going too far.")

The dependence of boatmen on tides must date back forever. Several other small craft owners and myself acknowledged it in a particular way in comparatively modern times – in Juneau in 1953. Our aim was to move our craft to summer fishing territory at Tee Harbor without going around Douglas Island.

How? "Over the bar" – meaning by an inside channel navigable only when the highest spring tide of the season made the water deep enough. We'd have to get up at 2:30 a.m. No use to argue. The tide was the boss. We got up at 2:30 a. m.

Tides have no sympathies. In Anchorage a boatman tied his small cruiser tight to a piling at high water and returned at low to find it hanging high for grinning photographers.

In Southeast, being a serf/fan of tides is the norm of the people. Tidebooks are standard household literature. Especially circled are the dates of the whopping spring runs.

"Ordinary" tides also get due respect, what with scientists pointing out that they continually enrich the intertidal zone, and boatmen using them to drydock and service their craft. But the big spring minus-4's-to-plus-18's remain the irresistibles, notably on the minus-4 end. Men, women, descendants, and family dogs go forth to exploit or inspect territory ever hidden otherwise.

The unveilings begin right in your harbor. The wharf becomes a skeleton on stilts, its pilings naked to their bases. The gangplank slopes near 45 degrees, and the *Z*'s of its rail-supports look more like *N*'s. Going down? With your heels gripping the cleats and the toes of your boots ramming your toenails into their roots (unless you decide to back down), you descend into the realm of the orange starfish on the gray harbor bottom. Coming back up? Like climbing a ladder. You're exercising your joyous subservience to tides.

4. A GAMUT OF EMOTIONS

Coming awake in the morning, you spend a few seconds adjusting to reality. What day is it? A workday or a weekend?... Where am I... What am I going to be doing today?

Where are you? Anchored in a nook in the forest – ah, yes, yes, that's it! About to spend another day doing nothing but fishing. Mooching commercially for salmon from a safe and comfortable little boat in the island-sheltered waters of Southeast.

The simple life, right? Wrong. Ups and downs, highs and lows, good sensations and bad. Sympathy, anger, anguish, apprehension, uncertainty. You'll be amused, frustrated, puzzled, disappointed, frightened, mystified, shocked, pleased, and delighted, and frequently more than one of these at once. Worst cases will leave you unable to say which of all was foremost.

Prime Time

Between the dark and the daylight,
When the night is beginning to lower,
Comes a pause in the day's occupation
That is known as the fishermen's hour.
– with apologies to Longfellow

It was the 21st of June. Slowly, expectantly, tensely, in the quiet light of evening, we mooched up on Three Kelp Hole at Grindall.

Day's end on Southeast water – there's nothing like it. You'd respond to it even if it weren't prime time for large and mighty kings. With diminishing daylight often comes a calm. The wind relaxes, the water flattens, the depths darken, the surface shines, the herring come up to plip and plop. The margin of the land exerts a pull upon the water. On the herring. On the salmon, found now in the shallower places. Especially, it seems, on the biggest kings, foraging in the shallowest. And on you the fisherman, too, naturally. But you pay the price. These twilights – they're too much. They'll charm you into exhaustion....

Three Kelp Hole hosted us possessively in the almost-dusk. Its dark green islet monitored us from one side, the larger and darker landmass of Grindall Point from the other. Rain had fallen recently and the jungles hanging above the rocks were dripping wet, highly inhospitable. A boat is a remarkable thing. In the absence of the boat what a lonely, heavy, difficult place this would be! Ah, but the boat is here. And you are in it....

The air was still, the water glassy, the world barely turning. Night comes slowly at this latitude, and the magic time lingers.

We crept past Three Kelp Hole, circled the islet, went through the point-islet pass, and along the kelp again, and once again. The herring plipped and plopped. Nothing else happened. Around again. Nothing. Charmed into exhaustion? I'll say. This June 21 day had been 20 hours long. So had the 19th and the 20th. So would be the 22nd and 23d. And 24th.

Mentally I dropped anchor in Coho Bight across the bay a mile away and squirmed into the sack, feeling its delightful caresses on my skin all over.

"Let's head her for the barn," I said.

Brother S is a persistent fisherman. Meaning stubborn, disagreeable, and unreasonable. He must always make one more pass. "One more pass," said he.

"We got up so long ago I cannot remember it," I said. "We've been to Vallenar today, and to town, and to Caamano, and to here. We are too old for this."

Brother S is 7 years older than I. "One more pass," he said.

This time a 44-pound king responded. And as the vagaries of the game go, of course, he took not S' spinner but mine.

I brought him home, and – "One more pass," I said. Unnecessarily. When you catch a fish, one more pass is standard procedure.

No salmon offered. We ran for Coho Bight, called up enough energy for grub, and squirmed into the sacks.

The Sleeping Giant of Chasina

The first chinook of the day set the pattern of unorthodoxy for the rest. He hit Brother S' spinner much too hard.

Not that we worried that S' shock-absorbing 20-pound outfit would come apart. Nor do we lack affection for chinooks that make us leap to our feet grabbing for our gear.

But theoretic, philosophic, and sentimental questions were arising, fodder for talk after we had netted the fish.

The time was the final week of June, when the largest chinooks usually begin showing. And we were working great chinook country, that stretch of Prince of Wales Island's east shore that we call Fishermen's Row. Ahead, southward beyond Skin Island, Chasina Point thrust its rocky snout into Clarence Strait.

We stowed the fish on ice, resumed our southward meandering, and got into the discussion. The main topic: salmon – the way they take a bait or lure, and the way they react after taking it, not to mention the way our heartbeats react to these ways of theirs.

Years of graphic examples issued forth from our accumulated stores. Most fish were more or less "normal," producing but a standard rush of adrenaline, but many were exceptions – nerve-frazzling oddballs.

Some of the oddballs you have encountered or will encounter elsewhere herein – mavericks that break the rules by foraging in the sunshine, skinny 11-pounders that strip out line like fat 50s, kings leaping like cohos, exceptions of all kinds.

As we reviewed the peculiarities of salmon here on Fishermen's Row this late-June day, an exceptional exception was still to come, a fish that planted himself firmly in our minds and brought along most of the topics we had reviewed.

The subjects arising included one you have argued, or at least examined, a few times. When a chinook takes a bait, should you set the hook or shouldn't you?

Brother S mostly doesn't think so, while I rather do. On this matter among others, we regularly have fun disagreeing. Now he used the fish in the box for support. "Why set the hook? He set it without any help."

"He was another exception," I said, "He had no business hitting so hard."

"No two are just alike," he said. "They're all exceptions, more or less. What do you mean, shouldn't have hit it so hard?"

"You know what I mean," I said, rising to the bait. At least you know why I say it – I've said it before. Your bait wasn't working right. It didn't fool the fish. He thought he had to kill it. He's supposed to think it's already disabled and all he has to do is enjoy eating it."

"I know, I know," he said. "That's your theory. It's probably wrong. At least, there's lots of chinooks that don't seem to believe it."

Brother S had me there... and anyway, if one spinner gets the "right" response, how do you hang on another that wiggles and limps just like it?

Still, the theory was one of my pets, a source of frustration from salmon that didn't buy it and of great satisfaction from those that did, and I would not then or now quite abandon it.

Our playful debate continued until additional oddball chinooks – not yet the super-oddball one – intervened.

"Wup! said S, and one, a 24-pounder, was on.

He bulldogged like a lingcod all the way to the boat, never making a single run.

Then – "Wup!" – and the next one, a 15-pound immature feeder of brilliant purple hue, hung down like a halibut and had to be pulled straight up.

The sky had been cloudy but now turned all blue and bright.

We were nearing the lee flank of Chasina, close enough to the point to see the foam that tailed out from the wash of the windward side when, once again – "Wup!"

The rod whipped once and the reel let out a single short snort. Thirty feet behind the boat, the fish that had smashed into the spinner on my short surface line went instantly high into the air. In a shower of sunny spray I could see my crescent-shaped sinker swapping ends beneath his huge form. Flailing mightily in that wet atmosphere, he splashed back down.

"Wow!" said S. "He's a monster. Better than 50 pounds."

The splash subsided. The great fish was still on my line, but, after that first savage strike and leap, was doing nothing.

In this crisis I used part of my brain briefly for the ridiculous purpose of being irritated. The fish had hit too hard.

"Bad bait," said S, helping not at all. How many fish, I thought, had swum up, looked at that spinner, and rejected it before this one got mad at it?

And now he wasn't mad. He was still there – a great weight on the line. But no action. Nothing. What the heck was going on?

I stopped pouting and started pulling.

The fish still did nothing. It was hard to get him moving – too heavy. But once in motion he came easier, like an endways log. I pulled him 20 of the 30 feet. Now I could see him, broad-backed and spotted, dead in the water.

Keep taking it easy, I whispered to myself. *B-Bring him easy and he'll keep coming.*

But zounds, what was going to happen when he got to the boat? The rod tip trembled, vibrating from my hands, right there in the tight air in front of me.

I kept slowly pulling, awaiting the surge and the flight. Brother S hadn't even picked up the net. I kept pulling. And staring. And pulling.

"That fish – he's asleep," said S.

"I... I – doubt it," I said.

Now he lay on his side beneath the gunwale in the light of day, shining and shimmering, four feet long.

Brother S held the net. "Now?"

I... I don't think so. No. Wait."...

So there we were, the three of us, a 50-pound-plus salmon on his side atop the water and two bug-eyed fishermen in the boat, one holding the pole and the other the net, and no one doing anything.

I added to my stress with the thought that a salmon of that size contains enough energy to travel 500 feet against a stiff drag. This one had it all still in him.

"He's a bomb," I muttered. "Touch him and he'll explode."

You've heard that the proper way to net a fish is headfirst; if he spooks he runs into the net. Don't you believe it. He can swap ends and miss the net entirely. Or snag the hooks on it and tear loose. If you've never lost a fish at the net, don't brag about it. It's a sign of inexperience. You do what the circumstances require.

"I'll... get him from behind," said S. "If he takes off you'll still have him hooked."

"Right," I said. "But go easy."

He sunk the net behind the salmon and moved it slowly forward. The hoop reached the fish's midsection, the bag billowing out behind, the fish's broad tail visible through it, nothing touching anywhere. Another six inches. Contact somewhere was imminent. Brother S didn't wait for it. He heaved.

The rim hit the fish behind the head, but the bulk of him was safely inside.

The energy was unleashed. The net bulged and twisted. Flying saltwater drenched us, as that lunker chinook finally came to life. What a terror he would have been on the line, in open water where he could do his strength justice. I almost felt like letting him out to demonstrate. Almost. But not quite....

Fifty-two pounds. Brother S says it is not exactly coincidence that this unorthodox lunker climaxed a day of oddballs otherwise – funny things often come in bunches. Or maybe it's just that this exception kept us from forgetting the rest. What I still don't know is whether my bait really wasn't working right.

The Redemption of Skin Island

As a serious Southeast fisherman you will likely become well acquainted with Skin Island Hole, a redoubtable salmon retreat in Fishermen's Row. The blessings will be mixed.

Skin Island is a low, spruce-grown rock half a mile long. On a chart of Southeast's tumultuous topography it is but one of the thousands of shapes in the water. But approach it by boat and it assumes a personality: it smites you with an attitude of gloom. It is spooky and foreboding. Even the overcast seems denser here.

Places of odd impact, both fascinating and depressing, lie here and there in Southeast. A cruise down Behm Canal northeast of Ketchikan gets you to one. It is a cousin to Telegraph Creek, not exceedingly far away in the wilds of British Columbia.

At Telegraph Creek, music is said to come out of the mountains. In Behm Canal, not tunes but voices reportedly filter down a wooded draw to the waterfront. (We went there to listen-in once, but no one was talking that day.)

You will find places in Southeast that beat Skin Island for scariness. One sits a short distance north of the Clover Pass entrance to Behm Canal. Here, among twisting channels between Tatoosh Rocks and Betton Island, cliffs and peaks have arranged themselves in such malice that the place is called "Gloom Alley" and "Death Row." In reinforcement of wickedness, a human skeleton reportedly was once found on one of the peaks.

To get back to Skin Island, despite its down side, it has always claimed at least one virtue: Run a few minutes from it and you reach secure and picturesque anchorage in Clover Bay.

Brother S and I first entered Skin's realm in 1955. We felt the gloom at once.

Brother S stood up, sat down, squirmed, fidgeted, and stood up again. "Something's weird around here," he said. "Whew!" I said. "Glad to hear it. I thought I was imagining things."

Fifty feet high among the boughs of a bushy spruce tree on the island, a dozen sea gulls sat, white against the green. "Never saw sea gulls in a bushy tree before," said S.

We went ashore and encountered a separate oddity – a hemlock tree, four inches in diameter, that wasn't sure which way was up. It had bent earthward and then swung up again, forming a complete loop. "Never saw a tree do that before," said S.

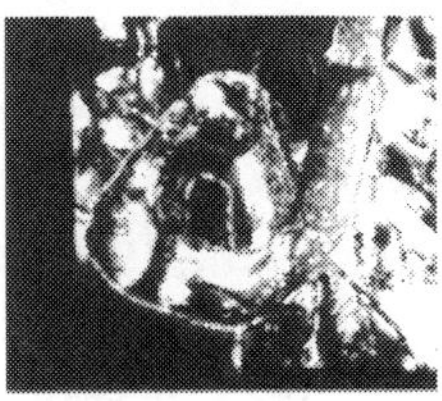

TWISTED TRUNK: "Never saw a tree do that before," said S.

We never have fashioned any link between these anomalies and the gloom. All we know is that they were there.

The Skin Hole was there, too, and although our own skins prickled, we dutifully fished it. Topographically it was not a hole at all, of course. But as a place to invite fish it was a dandy, an elongated lair for chinooks, a strung-out mess of submerged boulders and flowing kelp flanking the island – foul ground for fair, the kind of territory where, we'd learned, big spawners like to linger.

We fished it year after year, encouraged by good signs. Baitfish were regularly flipping, and the sea gulls – although we never saw them take to the spruce tree again – were usually in attendance. But we had barely enough success to keep coming. And always we fought the gloom.

June 1977. In one boat, we were mooching among the rocks.

"Good thing it's not rough in here," I grumbled. "We'd be bouncing on boulders. I still don't feel safe in this place."

"We'd better stick it out," S said. "This has to be big-fish water, especially at this time of day." It was 9 p. m.

In 20 feet of water alongside a boulder we said "Wup!" at once: two powerful fish, both solid on the hooks.

S' chinook headed east, mine west. How could we chase either without losing the other? We clamped down on both, stopped them in time, landed mine, pursued his in the diminishing gloom around two boulders and a kelp patch, and landed that one too, with a mighty thump inboard: 43 and 44 pounds.

Skin Island has seemed somewhat less hostile since. Death Row, though, probably must remain forever formidable, being in an area now closed to commercial fishing in chinook season. Even if a couple of 60-pounders capable of redeeming it should be lurking there, you can't go after them.

Money

You feel great. From the upstairs office you tromp briskly down the stairs to the plank deck and, through the fogs from the cold rooms and the aromas and the fish processing lines, stride happily out, primed for another trip.

What has brought on this euphoria? The papers in your hand. Ah, those papers.

The big pink one is your fish ticket or "pink slip" and is relatively detailed. By species, poundage and price-per-pound, it describes the catch you brought in.

The little yellow one is more brief and blunt. It says –

"Pay to the order of...."

Powerful, that yellow one. It justifies the whole series of adversities you've undergone to get it – the aches and pains, the long hours, the sting of cold salt spray in your eyes and cold fish slime in your cuts and scratches. That check – it's money.

Money. Success. Your first check swells you up, but so do the rest. Consider yourself progressing when first you are moved to establish a Southeast bank account, being uneasy about carrying the green stuff around.

Of course, you would not discredit the view of any friend or acquaintance who deems fishing its own reward. He makes a good case – "I don't need to catch anything. I just have to get out. Enjoy myself. Come back feeling good." But when pressed he will also admit... "Well, yes, that's true. If I've caught something I feel even better."

Where this fellow's reward is up one step, yours is up two. Money talks. In the "sports-commercial" concept of Alaska fishing, it talks loudly and legitimately.

Money is at the root of all your activity. You can't operate without it. Fuel, food, and moorage cost money. So do unlimited incidentals. Money goes out. Money must come in. Coming or going, money does what nothing else can.

Money, like lovely wives and old cronies, merits and receives affectionate forms of address. What's a big fish worth – so many dollars? ...Dollars? Plain old dollars? Too dull. Bucks, clams, rocks, smackers, simoleons.

Money will vie with fish for space in the records you keep. Newspaper column example:

> ...Brother S has agreed with a great idea I've had. This is that the fish we are catching are worth $6 apiece. This figure is an average of all – kings, cohos and pinks, lunkers and little ones alike.
>
> Every time we land one we say we are "six clams richer," regardless of kind or size. Brother S balked at the idea at first. Big kings are scarce, and he didn't like spending 20 minutes boating one worth $50 and getting only $6 credit. Neither

> would you. But how about six bucks each for little 75-cent pinks, coming aboard one after another from a big and hungry school?
>
> Cohoes, incidentally, are in the neutral zone, with an average value of almost exactly the specified six rocks.

Another time:

> We fished hard today but didn't catch much – didn't catch anything, in fact, until afternoon. We found a way to keep ourselves consoled, however. The gear we had in the water was worth only a few bucks, and our little trolling motor was burning less than a gallon of fuel an hour.
>
> "So we're sure not losing much," said S. "And as soon as we catch as much as one medium chinook we'll have a day's expenses."
>
> "Make it a big chinook," I said. "Two days."
>
> It was not long afterward that one of my rods nodded, nodded again, bounced up and down, and finally bent and stayed bent while the reel shrieked and shrank.
>
> "Three days, maybe," said S.
>
> Forty-four pounds. Brother S was right.

These scenarios, I concede, waive certain long-haul expenses. The annual bills begin accumulating well in advance of on-site operating expenses. Fishing-vessel registration costs $20 and the annual handtroll license $50 for residents and $150 for non-residents (late '90s figures).

You'll also spend more than $1000 traveling by land and water to and from the fishing grounds with your outfit. About $800 will go to the water part alone, ferry fares between Prince Rupert, B.C., and Ketchikan.

These are the minor amounts. You pay the big bite – although this one just once, not annually – for the right to fish at all. Under the Limited Entry law you get a permit only by buying it from someone else.

Permits have become valuable commodities. Some are sold directly by fishermen, others through enterprising brokers. In the mass-production fisheries such as seining, permits command prices running well into six figures. Your handtroll permit, the economy version compared to all the rest, set you back about $5000, per the 1998 market.

How long may it take you to get this back? It depends. Depends on many things – on the fickle fishing situation, on the weather, on the competition, and on your skill, luck, health, and persistence. Some days you may make hundreds of dollars. On a rare occasion you could possibly top a thousand.

A thousand, with little light tackle and a little light boat? True, you'll need multiple fish, and even if they're No. 1 Large Reds they'll not likely do the job by themselves. You'd best be getting big prices too, like those of 1988. That year, Large Reds climbed to $3.57 a pound at the buyer's dock. And for Christmas the mail brought a bonus that raised the per-pound rate to $4.45.

At this figure, half a dozen 40-pound No. 1 Large Reds would net you $1068....

No, we've never done it, but it's still possible....

And be not concerned about making too much money. The good news admitted by the Internal Revenue Service is that the fancy new motors and sophisticated electronic fish finders that you need to improve your fishing are, despite making it more fun, tax-depreciable. Big payoffs, meanwhile, join big fish as lasting season highlights.

The Stuff of Memories

We were troll-mooching off Caamano Point when Brother S tipped his head at our No. 2 Rock. "You keep telling me you'd like to fish from there," he said. "You'll never get a better chance than now."

The day was rainless, the water calm, the rising tide lapping gently at the steep offshore side of the rock. Close above the ripples lay the flat top, dry in the air, the safe and level platform of my plottings.

From the depths flanking this rock we had pulled fish when passing so close that we could clink a rod tip on its hard face. Why not mooch it from its own solid surface?

You have, of course, had a few experiences with best-laid plans. On the Southeast fishing grounds you'll have quite a few more. But hang in there! Get more than you bargained for....

I turned my "pilot's seat" at the kicker over to Brother S, and he put me off on the rock. I had yet to make my first cast when the bonus showed.

Under the title "Clear Water" some pages back I spoke of those rare occasions when remarkably transparent water yields views of subsurface action never witnessed ordinarily.

Now comes a modification. When conditions are otherwise optimum, total transparency becomes unnecessary. An advantageous viewpoint and a kind daylight that subdues reflection may be key factors. Whatever the secret, chance upon it and you steal front-row admission to underwater sights you've really no business seeing.

If you had no opportunity to commit larceny of this nature in your youth, or had it and didn't, start as soon as possible. Lifetimes are short and opportunities uncommon. The few pictures stolen will last and last.

Our accumulations of the scenic loot began in relatively lavish manner enabled by residency on pre-1930s Puget Sound. Roads penetrating this timbered country still had a long way to go. Commerce was yet dependent on the "mosquito fleet" of small boats serving the waterfront towns.

There being boats, there also were wharves and docks. And in those days, friend, when pollution was in infancy and marine and upland wildlife populations still rivaled human ones, these expanses of plank on piling were fishing platforms of the first water.

Southeast today revives the world we had, with enhancements: far greater distances between communities, in country much more rugged. By comparison, our lovable old wharves on the Sound included three reached easily by beachwalking from our family farm and several more within rowboat range.

Priceless were those wharves, hiding places for fish and vantage points for fishers. We had a routine: chip barnacles from the pilings to uncover pile worms, catch pogies on the worms, and use chunks of the pogies to lure rock cod (various rockfishes) hiding beneath the vantage points.

And in the evening, after all this, unleash the appetite of youth on white-flaked filets of cod sizzling under green flecks of parsley on the farmhouse table.

It was, as you see, a soul-satisfying regimen, every step of the way.

It was nevertheless subject to improvement, as Brother S detected one day while face-down on the dock, eyes fastened on his bait in the shadowy depths, tensed for the incredible thrill of actually watching a fish grab hold.

From beneath a shimmering cross-brace of timber, the biggest rock cod in Kitsap County hove into view.

"He didn't pay any attention to my piece of pogy," S told me. "He was chasing a pogy of his own. Caught him, too.... We have to start using live pogies for bait."

So we did, and they caught cod. And that is not all they caught. S was lowering one along the dock skirt when an 8-pound silver (coho) salmon grabbed it. We ate him. He was good, though no better than rock cod. But we had expanded our realm. We had caught a salmon.

Did all this history flash before my eyes the instant I looked over the side of No. 2 Rock and saw that 25-pound chinook salmon finning there, tail to a bump in the rock and nose to the current? I think so. And I guarantee that one other piece of old wharf history did.

The bump on the rock became a piling on one of the wharves. Tail to the piling and head to the current lay a cutthroat trout of 19-inch, two-pound proportions. Two adjacent pilings hosted comparable trout. They darted away occasionally, but then resumed their stations. I was there, unfortunately, on an errand. I had no fishing line.

On No. 2 Rock I had one. I put it in the water in front of the salmon. He darted away like those cutthroat trout and, unlike them, did not come back.

I caught no salmon from No. 2 Rock. But suppose I had. All that old wharf excitement and a fish to boot? Too much to ask.

Bear Scares

Imagine something like this:

A big Alaska king salmon is peeling your reel down. Your little boat is being blown toward the white lather of the waves slapping on the sloping rocks ashore. On the seaward side, the dorsal fin of a killer whale slices air, five feet high.

"That killer's hanging around too long," you tell your partner. "He bothers me. If we could get ashore through the wash I'd go in and stand on the rocks to land this fish."

"I doubt it," says the partner. "There's a bear in there."

Did I say, *imagine* all this? You might not have to. In a long Southeast career you could experience the real thing.

Now, we never actually did, quite. We just came very close.

Was it scary? I say yes. Brother S says bears "run away from people, and anyway the whale wasn't still close when the fish bit."

All right – the whale had moved out a little way. But he still was there. And so was the inshore wind ...and the rocks and the surf. And in the belt of tall grass between the breaking waves and the forest, so was the bear....

Now, admittedly he was a black bear, less fearsome than a grizzly or a brown – but no critter to mess around with, either....

And – "Wup!" said I, as my rod tip nodded and my reel snorted, and I grabbed the rod. But no contact. No fish.

And that was the difference. By the skin of a salmon's tooth we missed out on a story to add to our collection....

For some kind of emphases, the motor now sputtered, and I felt the wind give us an extra shoreward shove....

The motor never missed another beat (if in fact it ever really had missed one) and we went on by the bear. I kept a wary eye on him. "He's standing up," I said. "He's spotted us."

"Don't think so," said S. "Their eyes aren't good. I tell you, if he saw us he would take off.... But now, on this wind, he probably smells us. We're a fishing boat. We smell like fish."...

Not even during trips ashore, S contends, have we ever been in real danger from bears, "since we are still here." My problem is that I don't know this until afterwards.

And S can't claim total indifference. We both got our hearts in our mouths once, by jumping a black at zero range (as reported earlier) while taking our boat up a creek. And another time got a scare during a short walk in the woods of Cleveland Peninsula.

We went there for a reason on top of our urge of every few days to poke around inshore somewhere. "There's an abandoned antimony mine up in those woods," a Ketchikanite told us. "There's a plank walk leading to it. At least there used to be. I think it's still there."

We found the walk – an incongruity in a wild and silent realm – and began following it. We had gone a few hundred yards when my nerves went on a prickling spree. We were unarmed, and in bear country, and in cub season. Then came a noise unseen ahead, surely a sow brownie, with offspring – what else? – and we agreed in whispers to go backwards.

In the safety of the boat, Brother S said we should at least have had a look. "Don't bother a bear and she won't bother you," he said.

"How on that plank walk was she supposed to know we weren't bothering her?" I asked. Apologize for voting for withdrawal? Not to this day.

The plank walk bear may or may not have been real. Of the Smugglers Cove bear there was no doubt.

Smugglers Cove is a calm haven in the east shore of Cleveland Peninsula a few miles from plank walk. It first captured our fancy for a reason entirely unrelated to bears.

We were spending a night there in shelter from a storm when, on our little battery radio, we found ourselves tuning in loud and clear on far-flung lower-US stations from which we had never even got static before. Some kind of atmospheric anomaly, apparently.

On the day of the bear we never got interested in the radio. The only thing the two occasions had in common was the mean weather responsible for them.

On the south shore of Smugglers Cove is another cove, smaller and calmer than the main one, a haven in a haven.

Now the beginning of the bad news. This haven goes dry at low tide. The tide was on its way out.

Our boat at the time was a 20-foot plywood semi-vee cruiser, almost flat at the stern.

"So – let's just sit still and let the water go away," I said.

It was one of my less brilliant ideas. Boats are not made for sitting on tideflats.

"We won't be able to move," S said.

"Don't need to move," I said. "And we'll be upright and comfortable. The tide will be back long before the storm quits. Drop the hook."

We had a few fish in the box and were prosperous. At our little galley table we played rummy, and meanwhile settled on the flat.

The returning tide was not yet lapping at our chines when, down the shoreline 200 yards away, something black moved. Bear. Coming our way. *Never thought of that,* I mumbled to myself. "It's – it's a grizzly," I said to S.

" Nah," said S. "Too small. Black."

"He sees us," I said.... "Or smells us. Or... smells the salmon we've got with us. It's a good thing we don't have one out in the cleaning rack, where he'd be sure to smell it."

"Makes no difference," said S, "He'll get wind of those in the box anyway.... But he'll go on by."

The 200 yards shrunk. At 75 yards he was nearly abreast the boat. Between him and us the tideflat was a freeway.

"We're sitting ducks," I said.

Brother S pulled his .22 revolver from a side-pocket.

"You.... aren't going to shoot him with *that*, are you? I gasped.

"No, I'll just scare him if I have to," he said.

He didn't have to. Blackie ambled on down the shoreline, never even acknowledging that we were there....

Nevertheless, I have never since proposed sitting out storms on Alaska tideflats.

If you would argue that examples like those above do not justify being leery of bears, you sway me not. It is a lifelong sentiment. It got confirmed along the way by word that respect for the animals is standard attitude among the Alaska guides who hunt them. "Never sell them short," master guide Hal Waugh once told me. "Grizzlies, browns, Kodiaks, even blacks – they can all be dangerous." It was a comment comforting to recall.

Good Losers

A friend of ours, Chet Guasta by name, caught the biggest salmon ever taken on sports gear in Washington State, 71 pounds. Chet is unique. We haven't heard him say the biggest fish he ever hooked got away.

To all of you who do say so, we say your misfortune is modest. In Southeast we have lost the six biggest, at least.

The hollow lost-fish feel – all fishermen know it. "Having the big one get away is not the worst of it," says Brother S. "What's really bad is losing him without ever getting a look at him – not knowing what you lost."

Usually you will know whether it was a salmon, but you could be wrong. The tackle-buster you never saw may be a halibut, seal, sea lion, porpoise – Dall or harbor porpoise. We have snagged the former and mouth-hooked the latter.

Halibut take the bait on the bottom, where the moocher regularly puts it, and they make runs that can match a salmon's, usually back down to the bottom after you have hoisted them a ways, but also horizontally.

The realization that it is no salmon you have hooked may finally dawn only because the flatfish is a two- or three-hundred pounder that you simply can't pull up on mooching gear.

As for the lost salmon, I will not now sadden us both with accounts of the whole half dozen. But the biggest of all must here be honored.

July 18, 1981 –

We were working peacefully along the Vallenar shore when, near a fringe of kelp 100 yards ahead, Clarence Strait blew open. Out came 10,000 herring and, beneath their tails, a salmon of preposterous proportion. For 10 feet he planed along the surface before settling in his own wake.

"Look at him!" gasped S. "He's five feet long and a foot and a half deep! Biggest fish I ever saw!"

Possibly I had seen a bigger one once – in a picture in the museum in Juneau: the Petersburg world record, 126 pounds....

We crept within 100 feet, rods poised for casting.

"Get inside him," whispered S. "Give him room to run out with the line... away from that kelp. We'll get him... He'll hit a sinking bait."

Overconfident talk? Not necessarily. For all the water available to him, this greatest of all king salmon could be as catchable here as a cutthroat trout ravenously eating mosquitoes in the corner of a beaver pond. Where a million herring swim, the one spiraling down on the end of a fishing line is the one a feeding salmon grabs.

We cut the distance to 50 feet. Both of us were on our feet. We cast to both sides of the last point of attack and let the lines run.

"Mine's stopped! said S. "He's got it!"

I said nothing but I could feel everything, inside S' shoes. Felt the great fish monkeying with the bait. Felt him pause and consider. Felt two long, long yanks. Felt us hold the line tight. Felt heaviness. "He's on!"

Power unlimited was unleashed. The rod doubled and the star-drag reel yielded half its contents in a steady sizzle as that monster salmon ponderously swam seaward on the line.

He took the exact path we had planned for him. Hah, what wizardry! Through S' fingers I was still feeling everything. I manned the motor and followed, speeding up until the reel quit going backwards and we gained a few turns.

Now we were in charge. The bigger a salmon the more methodically he fights, usually. He is powerful but predictable. You can handle him. He doesn't jump, usually....

He went deep, turned up, kept coming. Fast. "He's gonna jump!" I said. But he didn't. The strain softened. Then – nothing. The line went lifeless.

We cranked furiously, trying to defeat the truth. No go. Through the water came the two bare hooks of the rig, bright and silvery and shining, showing us how fishless they were. Blank. Empty. Absolutely. Irreversibly. It was so simple. No shattered tackle, no spectacular leap. Simply, softly, the hooks pulled out. And so ended the encounter of a lifetime....

Acceptance took us awhile, but we worked at it. "Good thing we didn't get him," said S. "We'd have a record we couldn't break. Anyway, he just won the last round. We won every other one."

Mixed with enough discipline, the thought was a winner. We had been privileged to discover that great fish, watch him, thrill to him, stalk him, tempt him, hook him – yet leave him to propagate his kind somewhere in some spawning river. We wondered what river – the Skeena, in British Columbia, or the Stikine, up Wrangell way? the Columbia, far south? or the Unuk, well up in Behm Canal?

"I'll go for the Unuk," said S. "it's the closest, and maybe the wildest. A good place for him."

Shipwreck!

Battered and weary, shaken down by the pounding he had taken from the seas to get back home, handtroller Jim Anderson docked somberly at the Knudson Cove floats.

"Jensen will never make it," he told a waiting friend.

By "Jensen" he meant Brother S.

The drama Jim's declaration leads me now to recount is a reconstructed one. As has been my lot another time or two, I was south on less important business during the week of the crisis.

Jim had last seen Brother S off the Niblack shore down Clarence Strait from Caamano Point, a stretch unpeopled and unprotected, its few bays and beaches bare to the weather. What Brother S would "never make" was headway around the point into the lesser seas of the homeward stretch across Behm Canal.

It startles me now to realize how much time has passed since – in July 1964 – this struggle against nature was waged.

And nature has not changed since then. Be wary. What happened is waiting to happen again. Brother S' notes go this way:

> Monday, July 6. Too rough to get around point. Anchored for night in little bay.
>
> Tuesday, July 7. Terrible storm. Big blow and pouring rain. Anchor didn't hold. Pulled it early in morning and went into next bay – looked calm close in. Had trouble going in. Big swells. Wind. Tried anchor again – didn't hold. Fish net flew back and tangled in small motor. Anchor line fouled in big motor.
>
> Rode boat to shore and put bow out. Held her like this for six hours in heavy seas on incoming tide. No damage to boat but full of water.
>
> Wednesday, July 8. Cleaned out boat and held her again for several hours. Waves got so large, couldn't do much...sometimes swells would throw her 10-15 feet onto beach. She still stood all this until at extreme high tide some large pieces of drift were hurled against port side, breaking plywood hull 5-6 places. One hole amidships 1 foot diameter.... Slept on beach with fire.
>
> Thursday, July 9. Boat full of rocks and sand. Thought she was finished. Started to strip her and set up camp on beach until could attract some vessel to take me back. After cleaning out and examining boat, got idea of trying to patch her up

> enough to get back. Had can of epoxy aboard. Made patches from floorboards. Dried plywood with fire. Used nails, screws, plywood and epoxy to patch her up. Another record tide at night put me afloat. Anchored far enough out so would go dry but would be afloat on next incoming tide.
>
> Friday, July 10. Spent several hours packing all gear back to boat. Water smooth. Took off to get around Caamano to Smugglers Cove. Made it all right and anchored there. Wind kicked up a bit. Decided not to cross Behm until water was glassy. Everything in boat soaked, except some gear in bow. All foods except canned stuff ruined. Could get by a week if necessary. Have got gas lantern going and have dried out quite a bit of gear.
>
> Saturday, July 11. Now 6 a.m. Will make crossing soon if water smooth. Noon. Still too rough to cross. May have to wait another day. Can get by several more days with food on hand. Water never calm enough to cross. Maybe in morning. Spent day tying leaders, fixing more things in boat and doing some reading.
>
> Saturday, July 11. No, Sunday, July 12. Got mixed up on date. Water looks smooth enough. Started across at 11:30 a.m. Water almost smooth all the way. Pulled into Cove [moorage in Knudson Cove] at noon. Odin [Odin Jensen, owner of floats, no relation of ours] was ready to notify Coast Guard.

Odin's decision not to call for help earlier was a vote of confidence; Brother S would be able to take care of himself.

> Monday, July 13. Went to town. Felt good. Glad that I could get back all right after the frightful damages: Felt good that I could do it.... Went to post office. Got pleasant letters from home. Felt real good. Went to Beanery and had rib steak. Went to Bon Marche and got new hip boots. Burned others drying by fire on beach....

By July 15, Brother S had completed repairs on the plywood hull and the only shipwreck of our careers was history.

5. NEWS MAKES NAMES

On arriving back at the floats on a July evening in 1980 – one of years we fished from separate boats – I found Brother S already there. So were two tourists, standing and staring. Near their feet, S had adorned the planks with three big chinooks.

I joined the admiration society. "That biggest one – he must go 40 pounds," I said. "Where'd you get him?"

"Forty-two," he said. "Right in front of The Tree."

The onlookers swiveled their heads. Hemlocks and cedars loomed everywhere. The Tree?

You bet, *The Tree.* I knew the spot precisely....

Coining special terms of endearment like that one will be one more enjoyable element of your Southeast career.

Doing this is an acquired talent. Helpful for development are (1) the feel that your work is fun, (2) an eye for the gist of the scene or the action, and (3) a partner like Brother S.

Most of your names you will award to places – places hosting memorable events. But the events, too, will merit titles, and so will certain unforgettable fish.

Fishermen's Row

A favorite fishing stretch of ours runs along the east shore of Prince of Wales Island from Grindall Point to Chasina Point, a distance of about 15 miles as the crow flies and limitless as moochers wander. This whole works we call *Fishermen's Row.*

ON THE WAY: On a flat-water morning Brother S heads across Clarence Strait for Grindall Point and another day's work along Fishermen's Row.

Fishermen's Row is much too crooked to be called a row, but this fact was no factor when, back in about 1955, we named it for the succession of stops where we had anchored, explored, lunched, beachcombed, and fished.

Our usual first stop in The Row is a big-salmon hangout you have heard about elsewhere herein. Mention of it by our assigned title distinguishes it instantly from every other pinpoint on Fishermen's Row or elsewhere. The name was inspired naturally by the kelp triangle marking the spot: Three-kelp Hole.

To the west across a small bay from Three-kelp Hole, a rock-protected little bight is set into the forest. We have named this haven twice.

We acquainted ourselves with it well back in history, on the first or second of what would become dozens of days-long trips to Grindall. We came upon it at mid-afternoon of the first day while sounding the waters of the shoreline with our baits and sinkers. We had dual purposes: catch some fish if we could and meanwhile find a nighttime anchorage.

The bight, with its ample rock sides and water depth, filled the bill. "Looks as if it'll keep most any wind out," said S. At dusk we went in, slept soundly, and in the morning arose refreshed. On subsequent evenings and later trips, more nights of slumber followed. So did name No. 1: *Sleeping Cove.*

Name No. 2 owes its origin to an additional feature of Sleeping Cove: it is no distance from the fishing grounds. It is in the fishing grounds. You finish breakfast, stow the dishes, crank the kicker, hoist the anchor, and begin fishing. In that order.

One September we got a fish here without fishing. We had anchored for lunch. Two poles were propped in their holders and the baits hanging on the surface when a 15-pound coho grabbed one. Sleeping Cove came also to be *Coho Bight.*

Outward from Coho Bight (Sleeping Cove), we followed an embellishment of kelp along a low rock point extending toward a channel. The kelp stopped abruptly. "Must be deep water," I told myself. I was sinking my bait into the depths when – *Zinnnggg!* A king stormed off with it. I landed him among the other rocks and kelp bordering the channel westward.

In the ensuing days, months, and years, at this same point down the kelpline from Coho Bight, sinking baits have taken kings repeatedly.

In the absence of inclination to assign names to noteworthy places, we might have referred to this one as "the dropoff off the end of that point at Grindall where we catch fish on sinking bait." Too many words – the fewer the better.

But it was not until our second season, as I recall, that we backed into the fewer and better by failing to catch anything in nearby fishy-looking waters. "We better go back to *Sinking Bait Point,*" said S – and a title was born.

Sinking Bait Point. Whenever spoken thereafter those three words instantly painted a picture of the king-rich spot they honored. Exactly why and when we deleted one word escapes us today – but Sinking Bait Point has become simply *Sinking Point.*

Up until now we have called it that only to each other. Any stranger who heard us talking could be geologically misled. Sinking Point? Is it descending into the sea? Hardly. It is as solid as the rest of Southeast rock. But for economical communication between old fishing partners, it remains "Sinking Point" nonetheless.

Rough Country

The hazards of topography that lurk east and north of Sinking Point get worse west of it. You'd think the boulders and cliffs that front the forest should stop at the water. They don't. They merely go down and hide beneath it.

In the early years we had no electronic defenses against grabby rocks and kelp. Depth sounders, when we added them to our gear, didn't help much. The transducers are transom-mounted. When the pulse goes down to see what's under the stern, the bow can already be in trouble.

Anyway, to veer away from one seamount is to risk another. The number of hooks, lines, and sinkers we've lost to the submerged badlands west of Sinking Point is substantial. But then, so is the number of kings we have chased around and among those boulders.

For whatever reason, though, the title we gave the place short-changes the kings. What could have been something like "Chinook Rocks" or "The Big King Boulders" is instead simply *Foulground.*

There came one lunchtime, still in our early days, when I trolled in from Foulground past Sinking Point to drop anchor in Coho Bight ahead of Brother S. He was still out in Foulground.

A half-hour later he pulled in and tied up to me. "We've got to get this eating business over with," he said. "I've got something to show you."

We got the eating business over with and I followed him out. He trolled to one end of Foulground and paused, eyeing Grindall Island. "See the big rock this side of the island?" he asked. "I see it," I said. "To the left of it on the island there's a darker patch of rock," he said. "I see it," I said. "Okay, we're in the right spot," he said. "From here we go straight through Foulground – head right at that bump of rock beyond the far end – see it? – and it's clear sailing. I just finished going through twice – once shallow and once deeper – and never hung up once."

He led the way in the designated direction. I followed. We had gone a third of the way when my pole bent over. "Hung up!" I shouted. "Hung up, nothing!" he shouted back. "You've got a fish."

A fish it was. King salmon. He took me off course but after he was in the boat with me I got back on and, as S promised, went all the way through without a hangup.

What do we call this thin slice of clear water where you can trail your bait through Foulground unimpeded (except when kings charge out at it from the badlands on both sides)? *The Slot.*

We are old fishermen. Younger ones might have come up with something fancier, like being "In the Groove." "The Slot," though, serves handsomely.

Beyond the west end of Foulground the inshore rocks step up to cliffs fronting the hemlocks. One day in 1971 while mooching along together in one boat off the sheer face, we found ourselves under observation from above.

"Wolf," breathed S. "Can't be anything but a wolf. Black timber wolf."

Never had we seen a wolf before. Plenty of deer and a few black bear, but never a wolf.

As a noble wild creature this wolf blew his act. He was stark against a stormy sky, his fur blowing in the wind. But a classic pose? Hardly. His head and tail both hung down; he drooped.

We edged closer, peering. Momentarily he feigned dignity. His neck went rigid and he froze on us, like a setter on quail. But then he flubbed again; broke point. Broke point, turned tail, and crept beyond our view.

In no way could his performance have prevented the labeling of his lookout. It is *Wolf Rock* forevermore. And what is more, beneath it is *Wolf Rock Hole*, because from the cliffside waters to which he had drawn us, we immediately pulled a lovely chinook. More about Wolf Rock Hole later.

High Island and Co.

Four miles south of Foulground lies another assortment of spots Fishermen's Row has forced us to immortalize in language. Several islands mark the territory, the main ones being High and Patterson. To be working the water anywhere in the vicinity is to be "fishing High Island."...

Off High Island sits an islet and off the islet a smaller islet. The three are placed and sized like a man's body, head, and nose – although hardly so shaped. The anatomy is admittedly disconnected, too; the poor fellow has a channel in place of a neck and another channel isolating his nose. On the other hand, he is rich if a fisherman. Parts of him mark the most productive water we've found while "fishing High Island."...

Brother S put me off on the nose-islet with my camera; I would photograph our 20-foot cabin job trolling through the neck-channel.

TIGHT QUARTERS: Brother S trolls through the neck channel..

Meanwhile I would perform a totally unrelated function: check the quality of a patch of salal growing on that nose. Were the sprays salable? On Puget Sound we had earned money picking salal and huckleberry brush for market.

The first thing High Island through a viewfinder will do for you is make you lower the camera for a bare-eyed look. In a realm of otherwise wild and random island shapes, High Island is circular and symmetrical. It is half a mile or so in diameter and 400 feet high, a rock dome topped by at least enough soil to support a neat pincushion of conifers.

I got the picture and examined the salal. It was scrubby and useless. But it did yield a mink. He scurried from the salal down the rock bank toward water, stopped a moment to leer at me, and disappeared in a crevice.

The salal bushes now shook, and out jumped a deer, a tiny blacktail, with no horns but no fawn-spots either. She (he?) did not weigh over 30 pounds. She plunged into the channel and swam for the steep rock flank of High Island, 200 feet away. Brother S landed at our pickup point and we watched.

"She'll never climb up there," said S. "Too wet. Too slippery. Too steep – 45 degrees. "She'll have to come back."

You'd have agreed. In cleated rubber boots you have slipped and skidded helplessly on lesser slopes than High Island's.

Turn around and come back? Not that little deer. On her hard hooves she stepped easily up the wet rocks and into the browse at the edge of the timber....

"She did it," said S. "It's true. We just saw it."

From High Island's nose-islet a reef extends. When tidal level and surface conditions are favorable, it is easily visible from a boat. Over this reef, as reported earlier in these pages, we spent fascinating moments above a school of herring. So transparent was the water at the time, and so clear our view, that we could plainly see the black pupils of their eyes. And gradually we made out much more – the mouths of these herring opening and closing and their fins moving; and finally, nearby, two salmon lurking.

Although we tried (as also reported earlier) we couldn't make those salmon bite, and the standoff ended when something like a third salmon streaked in from the side and routed the first two.

We moved a short distance westward. I was bouncing my sinker off bottom in 30 feet of water when suddenly the bottom disappeared. Then it came back up. "It's a pit," I told S. I dropped into it again, and pulled out a king. And soon afterward, another. (And on later trips, still others).

We can give you no formula explaining fully why our little localities get dubbed as they do. No formula exists. None should. Formula or system would spoil spontaneity. Most names arise on impulse and they stick, whether or not the best upon reflection. Result: along Fishermen's Row, the body, head, and nose represented by High Island and its two islets get no recognition in our "fishing High Island" terminology. You'd think we'd talk of working the belly, ears, or nostrils. We don't. Our course begins at *The High Reef* and extends past *The Mink Salal* and over *The High Well* into *Little Deer Pass.*

Garbage Point

Prince of Wales Island, the backdrop of Fishermen's Row, is more than 100 miles long and 40 miles wide. To a fisherman wandering among the numerous and various smaller islands strung along it, the big one seems a mainland. Its interior is bumpy with mountains and its shoreline ragged with bays and snaky inlets reaching out of sight among the bumps.

Three miles south of High Island, across one of the inward ramblings of the land, the main shore reaches northeastward in a gracefully rounded point. The sculptor of this curve added a little island off it and gave us *Island Point.* (This name we did not originate. It was on the charts before we arrived).

Island Point's virtues include its proximity to safe anchorage in Trollers Cove and its appeal to king salmon of large size. The bad news is that it attracts garbage.

"Garbage," as you know, means any accumulation of seaweed or junk that fouls a fishing line. But off Island Point it also includes the real thing: milk cartons, plastic bags, grapefruit rinds. While there is less of the stuff than in the days before save-the-environment became a cry, you'll still be surprised at how much of unpopulated Alaska's floating junk is man-caused.

Untidiness aside, where the currents of Island Point put trash they also deposit the baitfish that bring the salmon. Big ones – on two occasions for us, 50-pounders.

One of these wore a seaweed mustache into the net. He helped originate the sophisticated way we describe our Island Point mooching – *Fishing the Garbage.*

The Three Trees

The tree which Brother S referred to back at the start of this chapter as *The Tree* was not the first tree we have honored by title. There have in fact been three Trees. But no confusion, since general location tells us which one is meant at the moment.

TREE ONE stands alone and unprotected on a rock at the entrance to Clover Pass. It is a Christmas-tree-sized hemlock, a lone little sentinel overlooking a fine fishing hole. From the start in the 1950s we made good catches beneath it.

"What I don't understand," said S then, "is how it can stay there through the winters. It's a remarkable tree."

Neither S nor I, however, was the person who actually named this tree. Brother A (Alfred), who also came north one year, was. Among big trees all around, he termed this little one *The Tree.*

In about 1982 Brother S was moved to new words about how tough life must be for The Tree, and so how tough The Tree must be..... "It still looks the same as it did 25 years ago," he said. "Hasn't grown. Still green and alive, though. It's a great tree."

TREE TWO, a cedar maybe 30 feet tall, got its recognition from us after it had died.

The year was about 1960. We had just made a double on kings over a reef off the Niblack shore and were fixing the spot in mind for future reference. "Can't miss it," said S. "Right on the reef and" – he pointed inshore – "right out from that tree hanging down from the side of the cliff."

Hanging down from the cliff, indeed. How it grew there, or why it died and fell, or how it then kept hanging remains a mystery but, barkless, bleached, and with a bare skeleton of curved branches, it hung there for more than 30 years, helping us get into fish-catching position over the reef. Each year for those 30 years, on our first trip to Niblack, we checked to see if it was still there and, incidentally, if kings still frequented its reef.

In 1996 an era ended. Time and the winter winds had prevailed. The Tree was gone.

TREE THREE had a shorter life. Yet an impressive one. I mentioned it earlier. It was a hemlock that had fallen outward at right angles to the land and lay afloat, roots still attached ashore. Around its tip we caught fish. But we admired it most for staying there all season despite Southeasters and 20-foot tides. It failed, however, to survive a winter; the next spring it had disappeared. Salmon no longer lurked where its tip had been, either.

White Lightning

Most of the names we've coined memorialize locations – Wolf Rock and Wolf Rock Hole for example, named because we saw a wolf and caught a fish. But we also have other excuses.

1981. In the calm warmth of a partly sunny afternoon, we were working Wolf Rock territory. The surface of Wolf Rock Hole, however, was cluttered with free-floating kelp, the flotsam of a spent storm. A 20-foot-wide belt of the stuff was strung against the cliff.

"Better try it" said S. "Kings hang under cover like that."

Never have we tangled with a wilder salmon than the one that laid hold under the trash. Against all the pressure I dared apply he streaked 100 yards toward Foulground, paused half a second and took another 50 yards. Only hot pursuit kept the reel half full.

When at last we closed the gap, the fish dove under the boat to leap high starboard while I was still facing port. I sunk the tip to clear the prop and come up where I needed to be.

"Crazy fish," said S. "Must be a white."

White-fleshed kings, we tell you, are more powerful and spirited than red. Fish scientists may not tell you so, but many a fisherman will. And in the boat this one proved a white indeed, all 48 pounds of him. He also was – and still is – *The Wolf Rock White.*

The Wolf Rock White made the case for another kind of special term. Before hoisting him to the cleaning rack, where he would hang over at both ends, Brother S tethered him with a rope through the gills.

"Can't risk letting a fish like that fall overboard while I'm working on him," S explains. All 30-pounders and up require this precaution and are therefore *Rope Fish.*

NOTE THE ROPE: "Can't risk letting him fall overboard," Brother S explains.

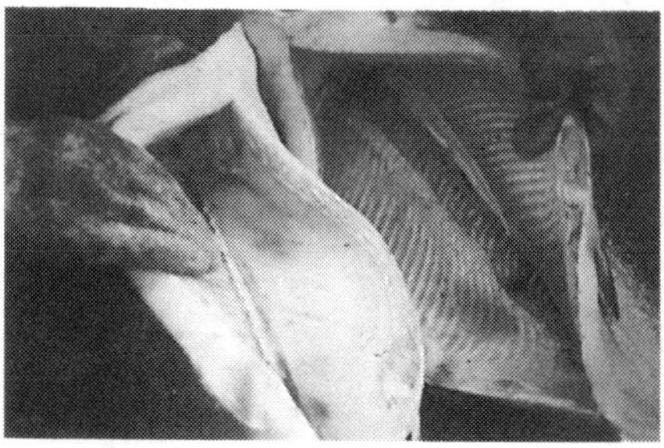

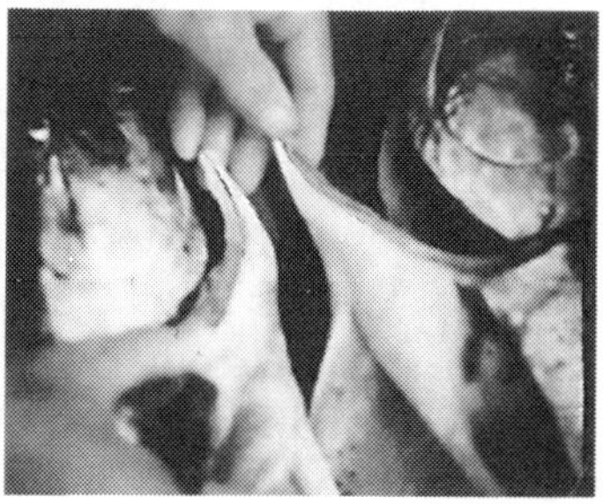

COLOR AND CLUE: Some chinooks have red flesh (dark, in these photos) and some white. Difference can be seen before fish is dressed. Photo at right shows where to look.

High Drama

With a great rustle and clatter, Brother S propelled himself upward from cockpit to gunwale to cabintop. He was 70 years old that year, 1979.

He made this maneuver in rainclothes and with the aid of only one arm, holding the other aloft. From this one a rod extended, whipping and warping to the antics of a very large king salmon.

Amazing, it is, what adrenaline accomplishes when crisis demands. And crisis did demand. The only way S could keep the salmon in the clear was to get elevation. He got it.

The scene was *No. 2 Caamano Rock.* (We also have Nos. 1 and 2 Rocks at several other places, and Nos. 3 and 4 at one). Number 2 Caamano Rock is a long, low one that thrusts its mostly flat top barely above water for a couple hundred feet near Caamano Light. It bespeaks raw environment. The seas of Clarence Strait pound upon it, substantial currents scour it, and underwater holes and shallows flank it: hazardous territory.

No. 2 Caamano Rock is thus a fine place for fish but a tough show for fishermen. Ordinarily, unless conditions are gentle enough to invite feel-fishing into its underwater foothills, we stay a respectful distance away, 50 feet or so.

On this occasion, wind-whipped waves were breaking white against it. But – "Look!" said S. "Herring jumping! Chased bait. We gotta get in there – get our gear in under those herring."

Near the rock, rebounding waves, random eddies, and thrusts of breeze made navigation tricky. I was making sure I had firm control before sinking my gear, when – "Wup!" said S. "Got one!"

So began the crisis. The fish, running shallow, streaked past the end of the rock and headed down the far side. A hundred feet of line was strung above the surface and sweeping sidewise, about to start rubbing on the rock.

That's when Brother S, at age 70, scaled the cabin.

What a sight a soaring eagle could relate! A salmon on one side of the rock, a cabintop fisherman on the other and the link between them whining in the breeze as the rock passed beneath.

When the water opened beyond, Brother S descended from his perch, braced himself at the gunwale, and brought aboard a 42-pound chinook.

"You're nimble for an old fellow," I said. "Made a lot of noise though. Damage yourself anywhere going up?"

"Damage myself going up? Don't remember going up. Must have, though, since had to come down."...

The fish joined others of fame as *The Cabin Climbing King.*

Fun and Games

For several minutes, while the current kept us in one position off a gravel beach, we watched a mink dig industriously among the pebbles. He paid us no attention; perhaps never detected our muffled underwater exhaust.

But when a salmon hit he did hear the reel scream. He jumped two feet, landed galloping, and was gone.

The salmon weighed 30 pounds. We are stuck with that weight, having recorded it. But the mink? "He leaped three feet," said S. "Went up four feet," I said. And for years thereafter, whenever we approached the gravel patch, Brother S grinned and remarked, "We're coming to the five-foot mink." Nonsense? Of course. But fun. And useful for communication besides.

But now, no way are we claiming this flair for foolishness as a virtue peculiar to us. Names and labels, nicknames and expressions, sobriquets and epithets. "Everybody has his own," says S.

The applications are unlimited and the benefits broad: memories preserved, sequences of events kept ever intact....

Apply your ingenuity. Build fodder for winter bullsessions. *The Tree, The Grant Island Log, The Alder Slide, The Otter Slide, Fur Farm Cove, The Black Sand, The Green Beach, The Wolf Rock White, The Great Pup Double.* Coin your own terms. Why say the fish bit "right in that place where the old cedar tree hangs down from the cliff on the Niblack shore of Clarence Strait" when all you need is "The Tree"?

A LONG, REMARKABLE STORY

While Brother S has been shipwrecked, I never have, although I did tip over in a dinghy once. This was in shallow water and no big deal, except that, in a camera in my pocket, was a picture of the biggest salmon I ever caught, 54 pounds.... Two months later in a darkroom I opened the corroded camera and developed the film and – should have been less pessimistic and more careful! The fish, except for the part of him that peeled off with the emulsion inside the camera, was still intact.... Next time I go swimming with a camera I will soak it again before removing the film. (Incidentally, fish look bigger with other fishermen. I'm 6-feet-7.)

6. ALMANAC

Count another blessing: the tilt of your favorite planet.

How does this aspect of your spinning, orbiting earth serve you on your Southeast fishing grounds? It triples the hours of daylight from winter to summer, thereby sending each phase of the unfolding season quick upon the last, never asking that you wait long.

There is also a down side. We have asserted that this fishing life of which we sing has scarcely changed in 40 years. The statement, as applied to the life, is correct. The daily joys and crises are as they were.

But with the brains and the brawn and the pressures of the fishing industry reaching levels not in prospect 40 years ago, or 20, or maybe even 10 or 5, something had to give. Mechanical and electronic wizardry have mandated stern measures to preserve the resource: curbs on fishing time.

You can still live the Southeast fishing life, but you must do it within trimmed schedules and under changing regulations. You've little chance of guessing what to expect this year or next.

In light of the uncertainties, the observations and events in the following pages reflect conditions as they have existed, rather than as they may or may not exist today.

Winter

If you have begun to feel you were born too late to get in on the ultimate fishing Alaska once offered, claim no exclusive right to mourning. Brother S and I were born too soon.

Juneau, 1952. The place was the office of C. Howard Baltzo, then assistant regional director of the U. S. Fish and Wildlife Service. The interview was unofficial, an aside from my newspapering job, but key parts of it stay quite well with me.

I was fresh from Washington State's Puget Sound, where the blackmouth (immature chinook salmon) of December and January adequately rewarded those of us willing to risk the rigors of pursuing them. Howard predicted that winter fishing – he was referring particularly to commercial trolling – would become increasingly important in Alaska.

Now, in the 1990s, Howard's prophecy is still coming true. The commercial troll catch of wild chinooks (distinguished from hatchery fish) has been growing rapidly.

How is it then, that we were born too soon? In the 1950s, when we were still sufficiently young and tough to pioneer in winter fishing, we were committed to other winter employment. By the time we retired from these justice-obstructing jobs and were available, our circulatory systems balked at frigid weather.

But you youngsters are in luck, born to be winter fisherman. The rules of the 1990s are on your side, offering chinook seasons much longer in winter than in summer. Adapt and cash in.

But be ready for anything. Fishing time will depend on rules and restrictions not to be predicted very far ahead.

As already acknowledged, we can offer no lore on Southeast winter water. But we do know a man who can, a hardy fisherman named Bob.

We also know something about chill factor. So do you. But in your case chill factor may combine only temperature and wind speed. In ours it adds another influence: the moist and heavy air of the North Pacific. You may nowhere find any scientific listing of wind chill as aggravated by this factor. You don't exactly need to. Your cinched-up pores will tell you. When the temperature descends much below freezing in these climes, it's cold.

It was in the summer of about 1980 that Bob briefed us on the trials and rewards of winter fishing. His boat was a 21-foot Glas-

Ply handtroller. Bob was approaching 50 at the time, young enough to do strenuous things and old enough to know what he could not do. He was of substantial build and in good shape. Besides fishing in the winter he ran a trapline, and he also packed out all the deer he needed for his freezers.

"The trapping and hunting are both easier than the fishing," he said.

"Meaning you get plenty of fish?" I asked.

"I wouldn't say plenty. Some. Half a dozen or so on a good day. But they're chinooks – that's what's here at that time of year – and a half a dozen mean 100 pounds or more. Enough to make it pay.

"They're not hard to find. They hang around in little sheltered places – places where the summer spawners, headed for their streams, don't stay long. In winter there are chinooks in the channel right in front of Ketchikan. Commercials can't fish there – it's closed. But a few sports fishermen do, and get fish without spending much time in the weather, minutes from home."

"Does freezing weather ever stop you?"

"Never has. Almost, though. The boat's got a tight cabin with a heater in it – but you can't run gear from there. You can't run it right with mittens, either.... They're all right for cranking up the gurdies but no good for handling leads with fish on them.

"But that isn't the first problem Breaking ice to get out of the harbor is. Noisy operation. Hard on the boat.

"One day last winter the temperature went down to about 12.

"Pretty good wind, too. Spray freezing on the wire line. No problem there – it cracked off going around the gurdy."

"Rough seas?" S asked.

"Somewhat. Sometimes. Not as bad as I've seen in open water in the summer. When you're close to the trees in bays and coves, the wind doesn't rile the water much – there's not much room for buildup. But it's hard on your hands and face."

"What shape are the fish in?"

"Prime. Fat. Top quality."

And, he said, top demand as well. Markets, restaurants, and private parties demanding all he can supply. Finally, therefore, top prices – "good for the bank account."

Old-fashioned bankers' hours, too. No daylight until 9 o'clock and darkness again at 3.

The Eve of Summer

In a Southeast mooching season, one of key dates for Brother S and me has always been May 30 – Memorial Day. And a Memorial Day that topped the rest was Memorial Day 1974.

After landing a 49-pound king early in the morning we spent six hours catching nothing. When a swirling wind hit us, Brother S rationalized, "There might be a storm coming and it's time for lunch anyway," and we dropped our hook in a nook in the forest.

There was no storm. With coffee mugs in our palms and hemlocks over our heads, we asked ourselves why, beyond being a worthy national holiday, Memorial Day was special.

Three reasons emerged. Two applied every year and the third arose this year of 1974.

Reason One was "Pop," our dad. Pop was born on May 30 exactly 103 years earlier – 1871 – in Veerst, Denmark.

As a strapping youth in 1891 Pop sailed to wild, unknown America. After a pause in Minnesota he pushed on to the wilder, greater unknown, Alaska. In two of the Great Land's great industries he labored, shoveling sluice-box ore for wages in Nome and fishing commercially for salmon out of Ketchikan.

What else he did in those years when the 1890s were becoming the 1900s, we do not know. But by 1905 he had hacked a farm from the forest outside the community of Manette on Puget Sound in Washington State, and meanwhile had been back to Denmark to return with a bride.

From this union came six children, of which Brother S was the third and I the fifth.

Pop had not forgotten Alaska. He enchanted us with salmon tales, including this one: His pay was a penny a fish, and when he caught a 100-pound king he refused to deliver such a specimen to market for one cent. He threw it back.

We never knew Pop to tell untruths. The possibility that he invented this story, or borrowed it, never occurred to us then, and we cannot grill him now. No matter. He had planted Southeast salmon in us. Of our failure to prove up until the 1950s we say more's the pity, and assign blame to the reasonable substitute Puget Sound provided until then. Meanwhile, we stick to Pop as Reason One for our attachment to May 30.

Reason Two was the strategic point in time symbolized by May 30 each year as the season unfolded, sharing what it held. The fifth month's closing days were portent days unmatched.

"It's not so much what we've got – it's what we're going to get," said S.

In other words, human chemistry got turned on. Hype internal mounted as this month wound down. Ahead to be savored were the longest days, the most and biggest kings, the runs of cohos and pinks. June, July, August, September.

Everything, almost, was still in the future. Big disappointments, if on the way, had not arrived. Official forecasts of runs didn't matter much. They weren't gospel. Bid the good ones right and the bad ones wrong. Replace them with your own. Spring forth from those waning days of May.

Reason Three, the reason that made the 1974 Memorial Day a special one, was a double reason. First, 1974 was our own anniversary: the 20th year since Brother S and I began mooching Southeast together. Second, this particular Memorial Day was the one on which we had caught this particular 49-pound salmon.

We talked about the 20 years and about the fish.

"We should have another 20 years ahead, before we can't handle the work," said S. I was happy to agree to the 20 more years.

The fish, however, revived an old disagreement.

What I had maintained was that with the passage of years the chinooks were getting smaller or arriving later, or both. "There used to be 50-pounders here in May," I had said. "No more."

"Wrong," said S. "Mere ups and downs. There'll be some."

It was an argument that the 49-pounder settled on May 30, 1974. Neither of us would let one pound spoil Reason Three.

Epilogue: Now, in the 1990s, our statistics do show smaller chinooks: no 49-pounders in May since 1974. Philosophy helps here. No big fish? No matter. No fishing. As of this writing, no general season enhances the month of May.

Now, therefore, to live the life as we have lived it in tradition, you must accommodate to schedule. Replacing those memorable Memorial Days with something else will not be easy, but worth trying. Anything close will be a winner.

The Second Half of June

Most of the time the rule is that when you and your partner hook salmon at the same time they will be of the same species.

In the second half of June the rule gets suspended.

"Wup!" said Brother S. "Fish on!"

"Wup!" said I. "Me too."

Brother S' salmon went straight down and didn't stop until bumping bottom. Thirty-pound king. Mine came straight up and didn't stop until three feet clear of the surface. Five-pound coho.

The action was a modest sample of what can happen in the second half of June. Brother S has a name for this special fishing period – "mixed bag days."

Being a salmon fisherman (fisherwoman, fisher), you already know about mixed bags. And therefore about mixed-bag suspense. Your pole bends, and with a "Wup!" you brace your knees and your jaw against the salmon that is bending it. Which of the five Pacific species will he prove to be?

Southeast's mixed bags, Brother S maintains, are more mixed than most. Especially in the second half of June.

Back in the 1960s we decided arbitrarily that the second half of June began on June 15, although on the calendar the right date is the 16th. But the 15th deserved the recognition. It was the day on which, by action of the fisheries powers-that-be, coho salmon annually joined chinooks as legal to catch commercially. Nature often cooperated as well, meanwhile impressing us with the great master timing system of the universe, by sending the first of the season's cohos to our lines exactly on that date.

Following the vigorous beginning each year, mixed bag time often accelerated. On the tails of the cohos came the pinks, small but scrappy, heading for Southeast's innumerable small streams to climax their short (two-year) cycle.

Chums and sockeyes, too, enriched the waters, and while these two species were more familiar to net fishermen than to trollers or moochers, we did get a few chums a year and once in a while a sockeye.

Hot fishing at mixed-bag time will give you a problem of communicating with yourself – that is, of writing the action into your logbook intelligibly enough so that afterwards you can decipher what you said.

“You need four hands and two brains,” says S. "When you’re catching the most fish, of course, is when you have to write the most words."

But why write words where letters will suffice? The next fish is out there waiting to bite. Quash the delay! Apply salmon shorthand.

SALMON SHORTHAND

SCIENTIFIC NAME	**COMMON NAME**	**LETTER**
Oncorhynchus tschawytscha	king, chinook	k
Oncorhynchus kisutch	coho, silver	c
Oncorhynchus gorbuscha	pink, humpback	p
Oncorhynchus nerka	sockeye, red	s
Oncorhynchus keta	dog, chum	d

In this neat system, a catch of three kings, nine cohos, and six pinks becomes, simply: 3k 9c 6p.

And we added improvements. Since salable kings can be either red (red-fleshed) or white and can weigh up to 50 pounds or more, we scrapped k in favor of r and w plus the fish's poundage. Catch a 40-pound red king? He’s an r40.

Good, but could be better. Which of us caught each r and w, and at what time of day? Piece of cake. We two fishermen were already shortened to S and E. And the hour and minute? Those are conventionally written in numbers anyway. So if Brother S got that 40-pound red king at a quarter after three in the afternoon – 315p Sr40.

In a maxed-out second half of June, fish shorthand shines. You’ve no seconds to waste. You get squeezed between two duties. One is to make sure you get the events of a hot mixed-bag bite recorded for reconstruction on some future winter night. The other is to keep fishing.

We consider it rather an oddity that, never yet, have we taken all five species of salmon on the same day. Four – yes. Once, in addition to c and p, we got both d and s – the two hard ones – and couldn’t land a k. Several times, only an s has been missing.

Mixed bag days get their special status not by happenstance. The solar system recycles itself in the second half of June. The orbiting earth and moon reach an alignment with the sun that maximizes forces. The exact time varies. The forces peak when the optimum alignment occurs in the middle of the period – around June 22.

On the fishing grounds look for:

– The year's longest days

– Extreme tides

– Salmon of any or all of the five Pacific species.

Whatever the possible connection, the weather also has been known to turn nice.

In any given year the odds are near prohibitive that all these phenomena could occur on exactly the same day. On the other hand, we have provided 40 years of opportunity.

June 22, 1978. Dawned now that once-in-40-years day:

– Fresh, clear air, purged of the storm that had blown us off the day before.

– Smooth water.

– The year's lowest tide, minus 4.3 at 9:34 a.m.

– Twenty hours of daylight.

– A mixed bag of proportions.

It was not one of those hot-bite days when fish stains can make the ledgers hard to decipher, but it was no day for complaining, either. We used salmon shorthand to record events then and we use it to present them now. With "how big?" and related information provided for each king and "how many?" for cohos and pinks, we give you this synopsis:

3a. Lines in water. Sky clear, beautiful day ahead. Feeling good. Couldn't fish in yesterday's wind so were in sacks by 8 o'clock, anchored in sheltered bight right off fishing grounds.

4a. Disturbance in water means salmon chasing herring. Good start. Er24 Sr26 1c.

5a. Water absolutely flat. Sky blue, wind calm. Two crows gabbling on sunny cliff ledges. Small deer eating kelp on 20-foot-wide strip of beach between two 20-foot-high boulders. Delightful world. "A day like this will make a young man of you," I say to S. Sw19 1c 1p.

7a. Tide going crazy, ripping at the lines. Sun hot, fishing not so hot. 1c.

10a. Tide gone slack. But what's this? A boulder we never knew about, sticking its snout out of water. "Can't understand it," says S. "Just goes to show you how much you don't know." Sr10 2c 1p.

1p. Fishing slow. Crazy currents on incoming tide worse than this morning's. Time out for lunch. 1c 1p.

4p. S gets a chum (d). Big one – 25-pounder. "One of a special strain," S says. "Twice average size." Tide slack again, 16.1 feet high. Fishing on top of the world.

7p. Temperature cooling off. Hit 77 degrees in afternoon. Good coho and pink bite for awhile. 4c 3p.

11p. Heading for anchorage. An Sr40 has climaxed the dusk. Dusk – that's when, S maintains, big ones bite... "not always, but often. Make sure you are fishing then." Sr40 1c.

Make sure you are fishing then? Good idea. But with limitations. It's the second half of June. The days are 20 hours long. You have to eat and sleep. And keep up the logbook. The requirement for discipline is excessive....

Watching S battle that final r40 finished me for fair. I had a smart remark to make but, lacking the energy to make it, thought it wearily to myself: *Days like these will make an old man of you.* Day's totals: 6k 10c 6p 1d.

Celebrating the Fourth

If big, fat kings and their court of cohos and pinks serve you royally in the second half of June, expect more regal treatment the week of July 4. If they don't, figure they're due to. Celebrate Independence Day. By now the chinook season is growing late, "and as we've seen year after year," says Brother S, "late is when the biggest ones show up."

S' comment agreed with one I had already heard from C. Howard Baltzo, fisheries biologist/administrator whom I mentioned earlier in connection with my Juneau newspapering job.

"After most of the kings are gone – that's when the 50-pounders-plus come in," said Howard. "We hear about it when they do, too – power trollers complaining on their radios of chinooks that wreck their gear. You can actually follow the migration of the fish by the sad stories of the fishermen."...

These oversized fish, he said, are dubbed "Point Colpoys" chinooks, because their migration route takes them past a landmark of that name, some 100 miles north and west of Ketchikan.

"The 126-pound chinook that's the largest on record was one of them," said Howard, referring to a monster caught in a trap near Petersburg in 1939.

While scouting for lone-wolf late-run 50-pound chinooks, you may instead run into thousands-strong early-run five-pound cohos – not that you'll mind. It is when you intercept both the king and cohos at once that the plot thickens. Question: When your partner hooks a fish, what should you do? (A) Pull your gear, to leave him a clear battlefield or (B) keep fishing?

Most sports fishermen (let us avoid all the possible "it depends" answers) probably opt for (A). Their sacrifice is noble.

As for us commercial types, we have come to prefer (B), having satisfied ourselves that an extra fish is more a bonus than tangled lines a risk....

From the vanguard of a surge of cohos we had boated two.

"Wup!" I said. "I've got another."

"Wup!" said S. "I've got – something. I've got – I've got a lot more than a coho. Big king.... He's taking off...."

"We gotta chase him?"

"Don't think so. I'll let you know."

While the chinook kept fighting the cohos kept biting. Before S pulled his soaker chineside and I paused to net it for him, I had brought in seven of them.

S' chinook weighed 47 pounds. My seven cohos totalled about 35.

As if prospects for bites like that one were not impetus enough to keep us on the fishing grounds and out of the city full time in early July, Ketchikan's Independence Day celebration was. Accumulating experience applied.

In the 1950s and 60s, when our precious Southeast fishing was limited to vacation time from other employment, we wasted none of it in town and never knew the trauma we were avoiding. Beginning in the 1970s, when we could avail ourselves of whole seasons and as much non-fishing leisure as we wished, we became more frequent urban visitors – until one July 4.

"Never did I see such patriotism," said S, as firecrackers rocked the rocky shores of Ketchikan and environs and reverberated off the rocky mountainsides. Half the city's dogs took to the woods and the other half cowered transfixed. By the time we escaped town we had sacrificed two-thirds of the age-damaged hearing we had arrived with.

"Never again," said S. "From now on we'll head out no later than July 2nd with enough provisions to stay in God's country until at least the 6th."

One day in 1979 this devotion to duty paid off. During an otherwise early-July slack period in the middle of a day, Brother S landed a 50-pound chinook and then, in the livelier evening, caught another the same size.

"We've never done that before," he said, a comment which led to talk of other things we haven't done: (1) catch all five Pacific species on the same day, and (2) land a 60-pound chinook.

That we'll ever accomplish either of these feats now seems unlikely. Though possible. At least one of our acquaintances has done the former and another the latter.

The fellow who got the big one was Russ Cowger, sports-commercial fisherman and one-time Clover Pass Resort owner who later was to become a charter boat skipper. When he caught the big one he was in a 14-foot rental boat – "rented" from his own place of business. In this craft He was fishing near us when we landed a 49-pound king. He pulled up alongside.

"That fish you just landed looked mighty big," he said. "Maybe a mate to this one I caught a while ago."

Not quite. His fish was a mammoth 64-pounder. It stuck out both ends of the wet gunny sack keeping it cool. It was a "hog" – a chinook with a forehead jutting sharply upward behind its snout, resembling the animal of the nickname.

After we pulled away Brother S used the occasion to review, as he loves to do, the imposing labels invented by fishermen for outsized fish, and especially for outsized chinooks. "Lunker, slug, slab, soaker, hog," he said. "They all sound big.... Which sounds biggest? Hog, I think. And hogs have that special shape besides. Makes 'em look bigger."

Hogs, slugs, slabs – whatever we call them, they help us handle Ketchikan's Fourth of July – make us concentrate on the fish and leave the land-based fireworks behind.

(Subsequent to the time we quit occupying Ketchikan during the Independence Day celebration, we're told, loud fireworks have been banned. We haven't risked going in to find out.)

Hard Labor

School kids out handtrolling ought not be able to influence two old fishermen like us, and until one Labor Day weekend they didn't. But during that three-day session a couple of them out-fished us, and we began paying heed.

They were different school kids from year to year, of course. What they had in common was an urge to make one more good catch before the return to classrooms curtailed their season.

We could not let them do this without a substantial effort of our own. In both camaraderie and competition we touched base with them from time to time during the summer. If they were going all-out for the Labor Day weekend, so were we. We made a tradition of loading up with supplies and, unless the weather was twice as bad as would ordinarily blow us off, spending Saturday, Sunday, and Labor Day in daylight-to-dark effort on the fishing grounds, preferably never returning to base camp.

"Labor Day weekend – we work hard those days," S confided to Eric, a new high-school graduate going South to college. "Have to. The year is about to end. Sure, we get big cohos later, but not for long. There's less than three weeks of season left."

For those three weeks the cohos usually cooperated. Whether by their own volition or by our extra effort, we found them in abundance and in fighting trim, 10- and 12-pounders prone to periods of rampage when they would bite anything.

We got so that we began anticipating Labor Day about the middle of July.

This is not to say that nothing happened in the six weeks between, although, as S commented to Eric, "Without the cohos nothing real great would," because the other big-money fish, the kings, "are mostly gone."

On Friday before Labor Day, S' sage words to Eric became proof that in the fishing life you can't be sure of anything.

We had joined Eric at the Clover Pass Resort gas pumps to fuel up for the big weekend upcoming when a guest of the resort approached.

"Think I can catch a chinook salmon out there today?" he asked. "Not likely," I told him. "It's strictly a coho show this time of year." So, of course, he went out and soon caught a 25-pound chinook. And we went out and caught another.

Exceptions like these, S says, "are all right when they happen, but they aren't absolutely necessary. The cohos can keep you interested enough by themselves."

Especially Labor Day cohos like those that Eric and other school kids helped us get primed for. On the Labor Day after the Friday of the two chinooks at Clover Pass, they were to inundate us at Caamano Point.

Eric and a partner said they were going to fish the west side of Gravina Island. S and I planned to cross Clarence Strait for three days and two nights along the crooked east shore of Prince of Wales Island, that stretch of fishing holes we refer to as "Fishermen's Row."

The Caamano cohos ruined our schedule....

Pausing at Caamano "to check it out on the way across" was a standard practice S always advocated. Check it out? We checked it out for three days.

When a coho bite is on, only a power troller can cash in fully.

And cash they do. "Sixteen lures out and cohos on all 16." Or sometimes it's "Sixteen lures out and cohos on 15 of them – I wonder what was wrong with that last one?"

Saturday and Sunday were good; Labor Day was better. When a bite started at Caamano Point, the best we could do was – four lines out and cohos on all four.

When a bite like this starts, you get only that first four-bagger. You aren't able to get four lines out again.

We spent the next two hours feverishly pulling fish, baiting hooks, wiping messes from hands, sticking hooks in fingers, untangling hooks from the landing net, and mopping the floorboards, laboring hard on Labor Day.

Get into one of these sustained riots and you won't forget it. If any respite stops the action all you have to do to start it again is to deploy your lines from pole holders and load your hands with gurry at the cleaning rack. Works almost every time. Keep your wiping rag handy.

The cohos were everywhere. In two hours we landed more than 35, averaging at least eight pounds dressed and 10 in the round. They were wild. Everything cohos do – hop, skip, jump, run – including running straight for us, they did.

One jumped over the boat, at the 7-foot width of the cockpit, and another jumped into it. Nothing new here – happens now and then. We know one fisherman who claims he has a way of *mak-*

ing cohos leap aboard – "Work fast, keep the slack out of the line, keep it tight while they're coming at you and just help them jump in. It's part of the show they like to put on."

AERIALIST: The antics of hooked cohos include jumping into the boat..

Our cohos performed another kind of act they "like to put on" when there are lots of them. When one was performing on my line a troupe of others was chasing him.

"Look at that!" wheezed S. "Two, four, six, eight of them!"

What these darting cohos are doing – sympathizing or harassing or something else, we don't know. But put another bait in the water, as S did now, and – bingo! Pay no attention to where you last saw the cohos doing their act. Sink the bait straight down and they'll find it.

Cohos – these grand fish accommodate themselves perfectly to the needs and desires of Southeast fisherman.

They don't wait for Labor Day – they're just growing bigger then. Anytime, as the summer unfolds, you can get accosted by schools of them. Nothing about them is ordinary, including their growth. They may go from about five pounds in June to 10 in August. Then they speed up. In one Labor Day catch we got two over 16 pounds. (For their size later in September see "Twenty Days," upcoming.)

Eric and his fellow student checked in with us in Knudson

Cove on Labor Day evening. They hadn't hit any such bonanza as we had. Still, they said, their catch "will at least buy our books for the first two quarters."

And we all bid one another farewell and good luck. Hard taskmasters but good fellows, those school kids.

Twenty Days Hath September

Twenty days hath September.

Or, rather, used to hath.

Nowadays it hath only as many as the fisheries commission decrees from day to day it can hath.

In former years the coho season started June 15 and ran through all of July and August and two thirds of September, ending September 20. That's when Brother S coined another of his terms. He labeled September a "20-day month."

But what a month. Outsized fish, vicious weather, strange surprises. Grab any of those 20 days you can get....

* * * * * * *

In a frigid, windless downpour we were trolling beneath Betton Island's western cliffs when one of our four rods bowed down and Brother S said "Wup!"

Behind us a fish came four feet up into the rain.

The rod doubled over again. The reel sang like a siren. Seventy-five yards of line vanished.

"Chinook!" I exclaimed. "King."

"Coho," said S. "Big one. King coho."

You've never heard of a king coho? No wonder. Brother S invented that term, too. When cohos go over about 15 pounds, he says, "they think they're kings, and act like it."

The fish made another long run and sounded, king salmon style, but finally came aboard. Coho.

Brother S hoisted him on the spring scale, jiggling the load for full benefit. Nineteen pounds. "Nice king coho," said S. "Unorthodox, though, right next to the cliff. They're supposed to stay way out." Indeed. But we had already been "out," and found nothing there but rain.

Unorthodoxy went further that day. It was a day of extreme tides, not a first choice of ours. Further, it was a day when more outsized cohos would launch their attacks from close in, not only from cliffside but from among shoreside rocks, another non-preference of theirs. Nice anomaly, though. As if the cohos of September's 20 days are not impressive enough in their standard irascible behavior, these added the extras.

Take nothing we are saying as suggesting that anything about fishing is ever safely predictable. All fish like to violate all the rules fishermen can ascribe to them.

Chinooks, for example, show up unannounced when they all are supposed to be long gone up the rivers. A handtrolling friend of ours caught a 50-pound milt-laden king in mid-September, two months behind time. We don't know where that fish thought he'd find a lady king to share parentage with.

The next day I landed a 40-pounder that got our attention because some of his scales came off easily in my hands. He was particularly slick to handle. He was very bright and shiny and he had a distinctive, sweetish odor.

"Dead giveaway," said S. "He's no ripe fish – he's a feeder. Wouldn't have spawned for at least another year. Wonder how big he would have been by then?"....

With the 19-pound king coho in the fishbox, we floundered on in the rain. We both stayed outside in it. Since running the kicker is my part of the partnership, I couldn't escape. Brother S could, but didn't. When the weather is good and the fishing bad, he'll often relax in the cabin for a spell. But in adversity he joins me in the cockpit. "Can't stand to see a grown man suffer by himself," he says.

Two features of September need emphasizing.

One is buckshot rain. If you are familiar with certain March and April showers on Puget Sound and have a good imagination, you can start to visualize what it's like. When the sea is calm, as now it was, pea-sized pellets raise multiple geysers that make the surface look like an electron-microscope photograph of something.

The other is that prized salmon of the late season, the coho or silver. In his final days he may also be called a hooknose silver, or simply a hooknose, this for the bent-down snout which the males especially develop at maturity. Ten pounds, 12, 15, 20, even 25 or more, this streamlined battler may weigh. "If he's hooknosed," says S, "He's as big as he's going to get."

Besides being at top poundage, the September coho is at top value per pound. Several times during the season, as the salmon grow and mature, the buyers typically boost the per-pound price.

If the fish has reached his prime, so have you, the fisherman. Hard as nails from the 20-hour king days, you can handle the shorter September ones in a walk – never lack for sleep and still never miss bite time.

In other words, at this time of year in Alaska, cohos and their enthusiasts get cycled into synchrony, coming together when both are in ideal phase of metabolism.

The cohos are completing a process of doubling their weight – at least – in three months.

Now, friend, they do not do this by fasting. Cohos-on-the-feed will bite on bait, spoons, plugs, flies, anything.

Their indignation at being deceived exceeds spectacular. Before our mooching days Brother S caught one on a dodger-and-herring rig and, before he could use the dodger again, had to reshape it. The coho had warped it tight over his back and was wearing it like a saddle....

Through the waterfalls flowing from our Sou'westers we began making out the Tatoosh Rocks, three grey versions of Seattle's Kingdome. The sea lions we'd seen airing themselves on the first rock on previous days were absent. They would have been wetter there than where they came from.

We were abreast of the first rock when, from wherever cohos get the "Now!" signal, they got it. Four of them – four at least – got it while swimming within range of our lines. What a chorus those reels played!

We landed three of the four, and they were the mere beginning. Twelve- and 15-pounders kept coming. Then in succession we got two 20-pounders-plus, two that together weighed 35 pounds dressed. Most hit on short lines in shallow water, often in 10 feet or less over a reef that runs out from the three main rocks. We stopped trolling and started casting, and then just sinking the lines; the cohos would take them before they hit bottom.

Once, after landing two fish, I was rebaiting one line while letting the other hang overboard with a battered bait still on it. A coho idled up alongside, watching it. "I think he wants it to move," whispered S. "Reach over and jiggle the pole – move the bait. He'll grab it and run if you do." I moved it. S was right.

For six hours, through two directions of tide and two inches of rain, while scrambling and "wupping!" to the shrieks, whines, and whistles of our reels, we kept dumping those cohos in the boat. A few times the fish paused to regroup – and to let S dress a few – but not for long. The air was growing colder and the daylight dimmer when the mysterious commandant of the school finally ordered fallback. In the fishbox were 36 of them, 323 pounds dressed.

In one three-hour session earlier in the season we'd made a catch of more numbers but less weight – 41 cohos averaging five pounds apiece. Only 41. With sports gear you have to spend time getting each one in, even when they're July size. In September you spend twice as much.

September buckshot complicates matters. You get wet. When I said in a previous report that a good set of rain gear will keep you dry in a buckshot rain, I neglected to emphasize that you need to sit as still as you can. In a September coho bite you don't sit still. The rain keeps looking for ways to get in, and when you change position, it finds them.

We can't tell you why our 36 big September cohos chose to be where they were. "Doesn't matter, though, I guess," said S as we were drying out in camp. "Still.... Well, do you think we're ever going to learn anything about this fishing business? Never? That's bad.... No, that's good. This way we won't miss out on the surprises."

7. LESSONS IN LORE

If some idea like feeding out line at the feel of a nibble occurs to you and you try it and it produces fish two times in three tries – bingo! Progress! But at a price. In the beginning you relied on a few firm and simple principles for filling the fishbox. After 50 years you'll have wrestled with hundreds – perfected some new, revised some old, re-discovered some forgotten, and debunked some once inviolate. And in the process, some will have become your mainstays.

Sinking into a System

Question: Of all the tricks you can pull on salmon to make them bite, which one works more often than any other? Answer: the one that triggers the action in every episode below:

Episode 1

My thoughts were not on fish – there weren't many around and I was capitalizing on the unlikelyhood of being interrupted by any. Trolling short lines from rods unattended in their holders, I was savoring the scenic Southeast solitude and the carefree feel – "fishing easy," as Brother S calls these departures from "work."

Scattered clouds moved slowly under the blue. Clarence Strait lay almost flat, barely wiggling me at the seat of my pants as I put-putted along, 100 feet offshore.

My inattention to salmon was abetted by the abundance of wildlife abroad. In one quarter-mile stretch I had logged two deer off one gunwale and a sampling of porpoises and killer whales beyond the other.

Joining these participants in the show of nature's creatures was one who was also an observer: a bald eagle perched on a cedar snagtop, a vantage point 10 times as advantageous as mine for eyes 10 times as sharp.

His head was cocked off my starboard bow, where high rocks dove to the mouth of a little stream, and another deer, maybe a 65-pounder, without horns, nibbled at kelp on a room-sized flat of gravel. Near the deer, a mink pawed at the base of a boulder.

The eagle sat. The deer kept nibbling. I continued slowly trolling. The mink saw me, flicked his tail, and scampered in a curve up into a salal patch.

On a ledge of cliff beyond, a crow sounded off. The eagle ignored him – never have seen an eagle catch a crow.

The sun came out, and my windshield reflected it playfully on the cliff. My indulgence in non-essentials peaked out. By shifting my weight and reversing and swinging the motor, I set about trying to put my spotlight of sunshine on the crow....

My control was bad and, as the boat slowed and stopped, got worse. But now the good news. The chinook that grabbed my port line and ran seaward weighed in at 33 pounds.

Episode 2

Once again, I was fishing easy, loafing along, my mind undisciplined, my whole wild universe unstressed, when came the event. Be not misled. Fishing easy is not the clue you are after.

This day's easy fishing was not as easy as some; rain squalls threatened – and now a maverick gust of wind plucked off my Sou'wester and tossed it overboard.

Reaction. Previous experience surfaced: Sou'westers sink, but not immediately. Move fast. Turn hard port, goose the motor, hoist the starboard line clear. Done. But when I was 10 feet short of my headgear a 25-pound chinook peeled out the port line. The rod hopped, the reel screamed, and the line went *wheeeshh*!. The fish ran outward and upward. He was going to jump. He did. At the corner of my eye, the Sou'wester went down.

Episode 3

My kicker had been acting up and, as I rounded a bluff in deep water, it lost one cylinder and quit. *Time for new spark plugs*, I said to myself. (Older outboard motors always needed new spark plugs). *Might as well put them in now. No need to pull the lines – plenty of water here.*

The plugs were in my tackle box under my seat. Getting them out and tilting the motor up took about 30 seconds. Installing the plugs would require maybe three minutes.

Those three minutes, though, had to start before ending. A 38-pound king on the port line fouled the schedule. With a dead motor and a very live fish, I spent 30 minutes on a landing job that should have taken only 10.

Episode 4

Fishing was good and I was expecting something to happen, when a chinook, with no advance nibbling or any such monkey business, laid into my starboard line full-bore, ripping off 50 yards, powering straight astern.

Got to go after him, I breathed. *Swing starboard. Won't cross the other line that way.*

Good strategy. By the time I was pointed the right direction another chinook was attaching himself to the port line. Both fish were hooked solidly and, after the stresses necessarily accompanying two-at-a-time, both were aboard.

Episode 5

The sun beamed brightly on a shimmering blue surface between me and the rocks, trees, and mountains. Mindful of theories that bright days push salmon down to 150 feet and more, I was drifting in 200. I made a short cast and, as I let the line run and began working up concentration for the feel of the 3-ounce sinker hitting bottom at 200 feet – it stopped abruptly at 50.

The first few times this happens, you will spend a second or so blinking at the rod tip or fumbling with the reel, wondering what's wrong. Whether you or the salmon figures it out first may not much matter. The battle is on.

My salmon came up from his 50 feet, churned the surface, dove to 200, ran along the bottom, and finally came up and curled into the net....

So now, of course, if you didn't already know the secret from experience before I tried to make a mystery of it, you know it now. Sinking baits take salmon like no other. And mind you, in the examples I've offered, the "technique" was unintended. Add intent and you get satisfaction in big chunks.

Perpendicular Trolling

The bottom was a full 150 feet down, and Brother S' feel for it as he let his line sink was further blunted by a strong current and a chop, and in any case he would rather work shallow water than deep.

"Missed it again – have to start all over," he complained, and began retrieving.

You know how it is. When you fail to find bottom at the expected depth, or aren't sure you found it, you're in a bind. To let out more line now is to invite birdnests. So retrieve and start over. It's no tragedy. It's all fishing.

Brother S' spinner was halfway back to the boat when a chinook ate it and, with help from the fisherman, completed the trip.

"Bait coming up produced again," he said, rigging a new spinner. "Still works almost as good as one going down."

Get that word "still." It tells you that this salmon had repeated a lesson others had taught. All your life, salmon will do this. Go back with us through four fishing phases (the first three are pre-Alaska):

Phase 1

We troll with a spoon in the least productive way possible: at a set depth and constant speed parallel to shore.

Phase 2

We turn to head back – and a chinook strikes. When this happens more than once – there's a reason! The spoon not only made a turn; its action changed and it went deeper.

Phase 3

We become moochers. Now we change direction and depth constantly, besides using the salmon's natural food as bait.

Phase 4

We go north, to Alaska. Here we make new discoveries and refinements, year after year. Some years stand out – 1978, for example. It was one of our best, both fish-wise and otherwise.

The "otherwise" included Roland and Roy.

R & R batched in an umbrella tent on a small flat where the vast forest of Cleveland Peninsula ends and the fishing grounds of Caamano Point begin, and they mooched these storm-prone, rainy waters commercially from a 14-foot open runabout. This tiny craft once had been a ragtop, but not now. Half the windshield was broken out, making any top unattachable. R and R braved the elements always, including when they crossed Behm Canal's four miles of unpredictable open water to civilization at Knudson Cove north of Ketchikan.

They had advantages. Sudden unfishable weather couldn't catch them more than a few minutes from camp. When they shoved off from shore they were fishing. They seldom mingled with the few other boats in the area, usually working – like us – nearer shore. Their boat barely moved as they huddled in their seats, intent on their work. And they caught fish.

All this we learned by observation and word-of-mouth. No one we knew had had direct contact with R & R. We came close once. We hooked a chinook, it ran toward their boat, and they came to life to clear the way for us.

"They're O. K.," said S.

As you can guess, our capacity for restraint was being taxed. Closer contact must be engineered. And it would be. But first, additional impetus arose, tied to a critical question among moochers: which bait is best?

Among the dozen or so sports fishermen and rod-and-reel handtrollers with whom we occasionally touched base, the choice was either whole herring or beheaded ones – "plugcuts." We alone used cut spinners, having discovered (Brother S did most of the discovering) that these thin, flexible filets of baitfish have several merits. They spin in a very slow current, they place minimum drag on the line, and they leave the hooks in the clear for hooking. They also may leak a trail of salmon-attracting scent in the water.

Their service as intermediary between R & R and us involved two enjoyable circumstances.

The first unfolded when Brother S salvaged an otherwise fishless day by hooking a brute of a chinook, landing it in sight of R & R, reading "49 pounds!" from the spring scale, and, at the cleaning rack, turning to show me two cut spinners.

"This one's mine," he said." I recognize it. But this one has to be – theirs."

The second happy happenstance occurred that evening when, at the same time we pulled into Knudson Cove, so did R and R. We didn't have to contact them – they contacted us, at our berth.

Roland, a compact man of 35 wearing bib overalls, Sou'wester, and shoe-pacs, was spokesman.

"Big one you got today," he said. "How big was he?"

"Here he is," said S, hoisting the fishbox lid. "And I think he's the same one you lost."

"Same one I lost? Well, I sure did lose one. And you think this is him?"

Brother S reached into his bait supply and, across the blade of his bait-cutting, fish-cleaning knife, displayed the second cut spinner. "This isn't mine," he said.

Roland peered. "Sure does look like one of mine."

Kinship was established. After supply trips into Ketchikan we held a session into the night, and bait talk led to larger elements of the trade, an exchange of high Phase 4 strategy. Join part of it in progress:

"....No argument. Fish parallel to the surface? We quit that years ago. The way to go is at right angles to the surface – up and down."

"And....why go parallel to shore? Right angles is better there, too – in and out from shore."

"Right again. Up and down and at the same time in and out. Cover all depths....lot's of fish in the shallow stuff sometimes."

"The way the shores and the bottoms are around here – crooked – sometimes we wind up going in and out anyway."

"Drift with the tide along shore. Same time go in and out and up and down. Cover everything...."

With that session, total right-angle fishing entrenched itself further as one of our stocks in trade. And so it remains. It can be far superior. Try it frequently.

Maverick!

At high noon on a day dazzling with sunshine I was trolling a spare rig with half-ounce sinker 20 feet behind the boat, fishing for a maverick chinook. So shallow was my spinner that, between each wave on the blue water, I could see it plainly, flashing silver in the sun.

I wasn't watching at the moment the maverick struck. I didn't need to be. A reel shrieking "maverick!" will wake the dead.

Before rescuing the flailing rod I cleared the field of my other outfits. With most fish I leave them in the water. But with mavericks they are likely to be in the way. The precaution proved wise. That maverick covered all the water around and under the boat before I got him inside it.

Mavericks by definition (mine) are rogues in their ranks. They hit savagely at the surface in the midday glare, sneering at the fish-behavior experts who assert that bright light drives them deep. Not only are they supposed to stay down – believers like an old power-trolling friend of mine will tell you why: "Their eyes are tilted upward. The light bothers them."

The theory sounds reasonable. But we have another: those few salmon that break the deep-water rule are all rampaging customers out to attack anything that moves. Mavericks.

Brother S, who is at least as fascinated as I by these surface marauders, thinks the maverick trait is "built into all salmon," but is seldom triggered. "It just gets set off once in awhile by some urge we don't know anything about."

"I hate to agree with you," I said., "but it may be so. If they acted this way all the time, they'd all have been caught by now."

Believe it. For additional evidence, look both to scientific studies of the remarkable sensors of salmon and to the impressions you accumulate on your own: when you put a bait in the water, every salmon within 100 yards probably knows it is there, and several may come and sniff at it before one grabs it. No such caution restrains a maverick.

Mavericks can get activated any time. They do not require direct sunshine. But you could swear they use it to prove maximum nonconformity.

What is or isn't a maverick by my definition can be a hard line to draw. Fish that come up because that is where the bait is, as at dawn and dusk, do not exactly qualify. On the other hand you do get them in the top water, so the excitement is much the same, whatever the words.

All old fishermen talk about mavericks, perhaps calling them something else. Or not calling them anything, making remarks instead like "that chinook was right on top in the mid-afternoon" or "all the fish seemed to be on the surface that day."

My personal treasury of mavericks includes one I featured in a column I sent from Southeast to The *Sun* in my home town of Bremerton, Washington. A "sunshine salmon" I called him, and on my return home a reader phoned me to report encountering "a whole school of sunshine salmon" at Sekiu in the Strait of Juan de Fuca, Washington State.

"Never saw anything like it," he said. "I was trolling deep when all these kings started going past me, sticking their fins out of water. I pulled up my line and one grabbed it. He was a battler. By the time I got him landed, the rest were gone."...

Another time, in Southeast, for several straight evenings one fisherman – fisherwoman, rather, she happened to be – arrived at the float with more kings than anyone else. Eventually, while transferring several big slabs from boat to wharf deck, she told her questioners, "They're right on top. If you can't lift your sinker out of water you are fishing too deep."

Inadvertently the lady made a case against trolling and for mooching. Trolling may mean uselessly working the empty space beneath the mavericks. Mooching, you test the top water often enough for discovery.

In honor of mavericks, S and I often fish a short line from an unattended rod propped at the stern. "Dummy," we call this "fisherman." Dummy has no intelligence or skill; he just sits and fishes for mavericks.

Dummy may catch nothing for days.

But when the mavericks get triggered he has been known to win our daily fishing contest.

Pound for pound, mavericks stand as the fiercest battlers of all salmon. There was the one that inhaled my spinner at full throttle and strung 100 yards of line along the surface without ever slowing down. What kind of monster was this? I got back some line but he took it away again. I remember thinking of a book I had read, saying that because of circulatory system design, salmon cannot sustain heavy exertion. This one could.

More revelation was to come. Fifty pounds? Stocky build, as the strongest fish seem usually to be? Neither. This maverick was a slim, trim, streamlined beauty – of barely 12 pounds.

Mavericks will pull other surprises on you. Encourage them, with that top-water gear. Don't expect ever to learn what pulls their trigger. Just be ready when it gets pulled. For multiple-maverick days be sure you're in good physical shape. Several mavericks in succession will tax your endurance.

Kings of the Shallows

Brother S swung his head in a slow arc at the shoreline greenery all around, meanwhile quietly lowering the fishbox lid on a chinook he had just landed and dressed.

"What I'd like to know," he said, "is why we catch this kind of fish every so often and still never know when we might."

"Wup!" I said. "We'll have to figure it out later – now I've got one." And I added mine to his.

By "this kind of fish," he meant those that by some urge or intent – or merely by wandering around for reasons perhaps not known even to them – make themselves available in flat-bottomed bays and along shallow shorelines. Powerful, exciting, memorable fish. You will find additional fond mention of them elsewhere in these pages. Too much cannot be said. When the bottom comes up until you can see the shadow of the woods upon it, don't hurry away. Scout out these places!

Shallow-water chinooks, says S, "Always give you something to remember them by." He is, of course, prejudiced by the wild surface runs big chinooks make when they can't sound.

And his sales talk goes much further.

"For one thing, when the bottom is right under the boat, you don't have to hunt through a whole lot of water for any fish that's in it. In the shallows he's right there."

How deep are shallows?

"Six feet is a good round number. Can be more. Or less. When the bottom is so close you start worrying about running into it, you can still catch salmon."

We began noting this in our first Southeast season. The location was off one "God's Pocket," in the Niblack shore of Cleveland Peninsula, a tiny bay with gravel beach behind a narrow gap in the rocks. We were in separate 16-foot runabouts. We landed, stretched legs, explored briefly in the mossy forest, drank from a little stream, and were ready to resume "work."

Brother S shoved off first. He turned the bow out into the narrow pass, tipped the outboard down, pulled it to life with the starting rope, and hung a bait over the stern. The water was not four feet deep. But a 25-pound chinook was there. It latched on, and it hauled him out into the strait before he hauled it aboard.

Brother S admitted no great surprise, even back then; he recalled shallow-water fish from our Puget Sound days. "Why, sure I thought I'd get one," he said. "I did put my line out, didn't I?"

And in recent years, he contends, the number of chinooks invading the shallows is growing. "Their habits are changing. We've always caught them in places like this. But now we're getting more than we used to."

We had just pulled three from sand-flat water an oar's length deep. "We fish these places more now," I said.

"That's because the fish are here. We've always done it. We just do it more now. The fish are changing."

"Maybe it's a case of different survival rates in different strains or runs of fish. A strain that likes shallow water is increasing."

"Changing habits or changing strains – for us it doesn't matter which, I guess."

We agreed on that point and one more – that chinooks in

shallow water can be hard on your heart.

"They scare you," I said. "I think they do it on purpose. All's quiet. Then one sneaks up to grab the bait and hightail away, like a kid committing some mischief and taking off laughing."

"It's a tight squeeze between the boat and the bottom," said S. "What do you expect them to do besides hit and run?"

"I like the mischief theory," I said.

We were, of course, not really disagreeing, only striving to do justice to these kings of the shallows.

You won't have to chase many of them over the flats, hoping to get your line back, before you too will be thinking up ways to express your sentiments.

Guide Dogs

Challenge: Find something good about Southeast dogfish. It can be done. We have done it. It isn't easy. From your first season to your last, dogfish will plague you.

"This small shark," the late career Southeast biologist Bob Baade says in his *Alaska Angler's Almanac,* "is attributed few virtues. Its generic name *Squalus* means 'dirty' or 'filthy' and its species name, *acanthias,* is the Greek word for thorn. It smells bad, too. This adds up to being a filthy thorny stinker, and if you don't believe it, ask any salmon angler who has been harassed by dogfish."

So what's good? To appreciate this you need first experience the bad. The dogfish will willingly demonstrate. After one day with them I sadly told my *Sun* readers:

> We are catching no fish. I mean, no salmon.
>
> But brother, the dogfish.
>
> For size, voracity, numbers, and razor-edged incisorship, the Alaska dogfish is one of which there is no whicher.
>
> Why, compared to these sharks, a Puget Sound dogfish gums off your leader after long and arduous labor. The Alaska monsters clip it so quickly your reel never even clicks. They probably don't spit out the hooks afterward – munch them up for dessert. Some will go five feet long.
>
> One of these, if you happen to hook him so he can't clamp his choppers on your line, will deceive you with salmon tactics. He'll bulldog, sound, run, and go airborne. Your elbows

> are galloping with his yanks on the rod and you think you've hung a giant king. Then he jumps and you get a good look. King? No. Knave.
>
> Once a foolhardy young salmon of about 12 pounds came up near the sharks to grab a bait which S had started lowering over the stern.
>
> "That salmon had to run for his life," says S. "And when he got tired and I was pulling him in, "two of those sharks came after him and were bumping at him with their noses. I thought I was going to have to beat them off with a gaff hook. Then after I netted my fish, those two and three more cruised along behind the boat and gave me the evil eye. I tell you, I'd hate to have fallen overboard about then."...

"Yes," Brother S assured a reader who later questioned us, "dogfish do attack salmon – especially schooled-up pinks. I think they figure that at five pounds the pinks are just bite-size."

And we still defend dogfish? You bet. To start appreciating the reason, go back to 1955.

The place was Grant Island, a tree-green jewel in Behm Canal on the way to the remote and renowned spawning grounds of the Unuk and Chickamin Rivers, and in its own right a place famous for salmon and infamous for dogfish. (Sports anglers today still wrest the occasional big chinook from the calm Grant Island shark waters, but commercials are restricted.)

We went there in the two 16-foot runabouts of our early years, rentals bearing numbers on their bows – affectionately "Ol' No. 5" and "Ol' No. 11." For me in Ol' No. 5, this was a return trip. I had paused here briefly while roaming Southeast alone two years earlier. For Brother S, all was new.

With eagles in its trees and deer on its shores, Grant Island welcomed us – esthetically, but not practically. How to fish this place? In Ol' No. 5 and Ol' No. 11 we drew gunwale to gunwale and were appraising the layout when – "What's that?" exclaimed S. A hundred yards distant, a patch of dimpling disturbed the flat blue surface. Then another patch.

"Herring," I said. "Something's after them."

"Salmon?"

"Not salmon. They're not jumping. Just balled up. Dogfish, maybe. But salmon still can be hanging around those balls."

We fired up the outboards and approached a ball apiece....

So compacted were the herring that those on top were standing on their tails on the writhing mass beneath, waving their heads in the air. They were 7- and 8-inchers. The rippled patches were maybe 20 feet in diameter.

I started lowering my bait nearby. Instantly several dogfish materialized and one snipped my leader. I tried again with the same result. Brother S was doing likewise.

We pulled abreast again, conferring while a ball of herring tinkled nearby. The stately hemlocks of Grant Island looked on noncommittally. Brother S had no saving proposal. In those days he still knew less about things like this than I did.

Neither of us was yet willing to rummage in our tackle boxes for the spoons and plugs of pre-mooching careers. Nor was switching likely to help, with these dogfish. "I think they'll eat anything," S said.

(Incidentally, you can put no stock in the oft-told tale that all sharks must turn on their sides to attack. These stayed upright from start to finish. See also pages 144-45.)

"The day is too bright," I mumbled wisely. "The kings must be here. But they aren't coming up. And we can't get down. Down? Hey! Let's move out, get down, and come in under."

A hundred feet away my bait went down without impediment. I felt it hit bottom, raised it 10 feet....

A nibble. And a tug.

Salmon. A 30-pound king glistening inside Ol' No. 5....

Do not expect perfection. Dogfish and salmon do not agree on the boundary between them. I started with 50 leaders and by nightfall was running low. But six kings justified the tradeoff.

The bait balls frequently went out of sight but then surfaced somewhere else. Like ancient mariners we stood to shade our eyes and scan the waters for dimpled patches. We caught salmon and sacrificed leaders until, three days later, the patches grew mysteriously absent and the whole show stopped....

For the next 12 years, while repeatedly relearning that, as S puts it, "dogfish mean baitfish and baitfish mean salmon," we never again saw anything quite like those Grant Island balls of 1955. Then came 1967 and the men from Minnesota.

We met them when, returning from three days on the far side of Clarence Strait, we stopped to refuel at the first outpost of civilization, the resort at Clover Pass. One of them told a sad tale.

"Herring – there were big balls of them, squirming on the surface, with a lot of big dogfish chasing them up. Don't know whether there were any salmon or not – the dogfish kept taking our baits and hooks."

I sneaked a look at Brother S. He was cautiously intent.

"Dogfish – they ruin everything," I said, carefully. "Was this down ... at Grant Island?"

"No, at what's called Tatoosh Rocks. Inside the rocks."

The picture grew doubly delightful: Herring balls rippling the surface. Inside the rocks. Smooth water. The rippled patches would be easy to see.

And still further impact: he described a part of the area more inside than the rest, with islands, reefs, rocks, and pinnacles, interspersed with deep water – a picture place for herring balls.

(The locality has yet another basis for intrigue, this one of foreboding kind; see "Death Row," page 75).

Soon after daybreak we arrived, with plenty of leaders and a plan: put our 1955-found skills to work on 1967 herring balls.

The rippled patches were there, and so were the dogfish and the salmon. The Minnesotans were not. We lost 30 or more leaders. But we caught five beautiful kings.

Lesser episodes of the rippling herring followed in 1978 and 1983. We can't predict the next one. But anytime the dogfish are thick, remember: until you have lost at least 20 leaders, don't quit looking for salmon.

Fringe Benefits

We were loading the boat for another trip when an aspiring young fisherman came boldly down the gangplank and said, "I hear you guys always find salmon. Tell me where."

"Well-l-l," said Brother S, swinging an arm in a slow arc covering all waters and shores everywhere, "if a salmon swims down he gets stopped by the bottom and if he swims up he gets stopped by the top. If he swims sideways he gets stopped by the sides. Where you find him is where he gets stopped."

Boundary Theory, call it. Whether the young fellow did or did not believe it, both Brother S and I do, and so can you. Boundary theory affects all fishermen, whether or not they so call it.

Before the banning of Southeast's fish traps in the late 1950s, boundary theory paid off lavishly in special application. Who could improve on a trap lead as a boundary? Hook-and-line fishermen made out like bandits at these artificial stopping places, not necessarily with the blessings of the trap owners.

But weep not for the passing of the traps. Apply boundary theory to the natural boundaries remaining – the bottom, the top, and the sides.

Of these, the bottom, in good old mooching tradition, stands as No. 1. From the inception of the mooching craft (unless someone can prove otherwise) the strategy was to explore the bottom: travel slowly and let the sinker fall freely. When you feel it hit, lift off. Meanwhile keep your senses turned on. Be a feel-fisherman. Ride the contours. Plumb the lie holes. Communicate with them through that thin, rubbery link of nylon. The first few salmon you catch this way will hook you good. My first ones, Puget Sound blackmouth (immature chinooks) in the 15-pound range, made a bottom believer of me.

When you probe that bottom boundary you're probing king salmon country – and this despite the fact that your bait doesn't stay near the bottom for long after you've free-spooled it down; it works up to trolling depth. Often the salmon hits the bait seconds after the sinker hits bottom. "If he's not already there, he follows it there," says S. "He's a bottom boundary fish."

Bottom boundary. Among moochers generally, it probably remains first choice. But it surely has competition.

You've heard of sports fishermen who thwart the law (you've never done it yourself) by sneaking an illegal extra line into the water, tight to the stern where the fisheries officer can't see it. For Alaska kelpers, extra lines have an extra virtue – they're legal. S and I have a special system. Instead of sending the line straight down with a heavy sinker, we fish the surface.

Brother S says I nevertheless still spend too much time at the bottom boundary "when you really knew long ago that the top can be just as good." He mentions an event on Puget Sound in the early 30s, well before our mooching days. By some hunch or obstinacy now forgotten, I hung a No. 4½ half-nickel, half-brass McMahon spoon with no sinker 30 feet behind the boat – and a chinook immediately hooked himself violently upon it. (If you try this, use a rudder to keep the line from twisting, which it somehow does even if the spoon is only wobbling, not rotating.)

The fish I caught that time was only an 8-pounder, tiny compared to those since encountered in Southeast, but an advance version of them nonetheless.

Brother S periodically recalls that fish of long ago to support his allegiance to the top boundary. "You spend too much time getting to the bottom and back," he tells me. "With short lines on top you can get your work done faster."

He has a confederate in his crusade: namely, that aerial acrobat of salmon, *Oncorhynchus kisutch*, the coho or silver. Cohos may completely shun both the bottom and the sides and crowd the top. Where the land is a mile or more away and the bottom hundreds of fathoms down, you'll find them, thick, ravenous, and rambunctious, smashing your gear on the surface. And when you hook one he often goes higher yet, several feet into the air.

Kings, too, have their days on top, as reported frequently herein. On a bright day in June I watched Ken, a power-trolling acquaintance, catch chinooks while I didn't. His 50-pound sinkers misled me. "The fish were right on the surface," he confided as we admired his catch at our anchorage that night. "Every one of those fish hit on my top lead."

It hadn't been my lucky day. I'd have been clued-in if just once I had contacted one of those hungry top-water fish while letting my gear down. Instead I had found all the spaces between them.

Next morning was too late. The surface fish were gone.

Top boundary, bottom boundary, side boundaries – S and I were approaching a pass between two islands, one of many such places in Southeast. The land on both sides was closing in and the bottom was coming up.

"Fish country," said S. "It's about time for a – Wup! There he is! But – nope. Didn't get hooked. Keep going."...

Nothing more happened. But look at the odds. Capitalize on them. Look for the narrow passes, the guts, the squeezed-together places in the migration routes. In these bottlenecks, salmon from cubic miles of water funnel into spaces better measured in yards.

The best fishing, says S, is "in the taper," where inbound salmon linger. "They go right through the narrow part. But in the taper before they get there – that's where you'll connect."

As if these places where the convergence of top, bottom and

sides do not make the fishing good enough, Brother S offers a way of making it better. Do it at daybreak, or better still at dusk, during that magic hour when the watery old planet is completing one more turn. Now you fish nearer than ever to the land, where the productive top and bottom boundaries are closest together. Be ready! Here in the shallows where the dark images of the spruces fall, marine life responds to age-old urges. Baitfish come up to plip and plop, and salmon move in on the prowl. Big ones. Engage one of them with mooching gear at this late hour and you face at least one extenuating circumstance. You'll have to land him in the dark.

Hole Hopping

If you were lucky enough to be born in hinterland country in times tough enough to make you appreciate it, and likely even if you weren't, rejoice at having landed in Southeast. The urge that brought you to these saltwaterways for salmon also sent you up creeks for trout, starting at about age 6.

Creek trout fishing – brushing the springtime greenery, treading the quiet moss, absorbing the music of the water, holding to an alder limb to lean far out and work a worm into a lunker's hole – this was an avocation not second to many.

To the possible occupants of the hole, a helpful general rule applied: big fish bite first. This kind of cooperation encouraged you to hole-hop, for quick coverage of your route.

Salmon, like trout, have holes, so you can also hole-hop for them – "go troutfishing for salmon," in Brother S' words.

Salmon holes aren't necessarily holes, as holes go; they are anywhere you catch salmon. Some are in likely places, like behind points or near kelp beds. Others are less obvious, but the salmon know where they are. Finding one by some minor indicator – tiderip, shady spot, underwater obstacle, hunch – gratifies the urge while feeding your ego.

Hole-hopping strategy is simple. A migrating fish comes along to occupy a hole and you then come along to fish it.

Optimum hole-hopping conditions include water calm enough for fast travel, a sufficient scarcity of fishermen so that your bait is the first one in the water for awhile, and the timing to arrive soon after the fish. Southeast mooching supreme!

We once held council on the question when and why we first decided to "troutfish" for salmon. Brother S searched his soul. "When did we make it into an idea – didn't just change locations because we weren't catching anything?... No, I can't remember."

Neither could I. But we both remembered "a lot of times since." An early example, as early examples do, still stands out.

It was a rare day in June 1957, a day to bask in, fish or no fish, a day of emergence from a week of foul, almost unfishable weather, a day to release the urge, a day as grand as an evil week of downpours and Southeasters could make it: a day of clean washed air, total visibility, blue sky with a few billowy white non-rain clouds, eagles silently soaring, blue water rippling, and Dall porpoises cavorting in Behm Canal off Tatoosh Rocks.

In the two 16-foot outboards we then were using, Brother S and I were trolling near the rocks. In response to the happy new weather, we gestured at each other in silly ways and mopped our brows in the mock sweat of 60-degree sunshine, exulting in our clean, wide world of freedom. Fish? It was mid-morning and Tatoosh hadn't yielded a one.

We squinted westward some four miles across the canal to Cleveland Peninsula. Bond Bay lay in an indentation there, and on its right the rocks and spruces extended toward us.

"North Point," said Brother S. "Trout hole."

"Let's go," I said.

You have deduced that, by the nature of hole-hopping, two people in two boats shouldn't be doing it in the same place at the same time. I hear you. On the other hand, who says practicality can't yield on occasion?

Brother S' boat was about a knot faster than mine. Three minutes ahead of me, he squatted to a halt. While I closed the gap as runner-up I watched him pick up a rod, stand up, check bait action, sit down – and stand up again in battle....

"Nice trout," he called to me. "Forty pounds, I think."

I gave him room to land it and he did: 43 pounds. He sunk his gear again, yelled "Wup!" and – 46 pounds.

Nothing more. North Point Hole was empty. "Green Beach," said S, and we took off down Behm Canal for that hole, which fronts a place where rocky shores give way to a small, gentle slope of crushed greenish slate. Thirty-two pounds. Mine, this time.

Return to North Point. Nothing. On across Bond Bay to South Point; one fish. In the evening, back to Tatoosh and one more....

In the boats of those days we did our hole-hopping at about 12 knots, a figure since upped to more than 20, the better to respond to urges, conditions, and hunches.

Hunches may pay off where there's in fact no hole. "Salmon follow boats," says one commercial troller. "I stop and drop my gear once in awhile, just in case."

Hole-hopping does defy the ancient truth that you don't catch anything unless your gear is in the water. But how many times have you tied into a salmon the first time you lowered your line and caught none thereafter? Stowing gear and running to a hole a mile away takes maybe five minutes and a quick test another five. The math comes to six tries an hour. You don't need a very high success rate to succeed.

So hole-hop. To all the places you know to be good add some you don't, and reap a double-barreled payoff: a fish plus a new hole to hop to.

King Kelp

Close to the kelp of Vallenar Point, a school of herring leaped in flight from a salmon. We didn't actually see the salmon, not at that instant anyway. Awhile later we did see him, safely in the boat, one more kelpside prize.

Kelp! Line-fouling, nerve-wracking, exasperating kelp. Fishy, helpful kelp. Kelp will ensnarl itself inextricably with your most breathless salmon stories.

Kelp! Remarkable kelp! As beachcombing schoolkids we made megaphones, snorkles, and blacksnakes of it, and in junior high class heard that it is rich in iodine but that no iodine could be found in sea water. Intriguing, mysterious kelp!

Kelp! The affection you will develop for it would shock the experts of old. Hear the words of one, named Andy, sharing the conventional wisdom of the day with two eager upstarts – Brother S and me – on Puget Sound about 1925:

"Kelp? Don't try to fish for salmon around that stuff. You'll get fouled up in it. Oh, you can catch rock cod in it, with a line heavy enough to pull them right up – good eating, too. But salmon? Stay in the open water."

Years later, on a day when we had just plucked two kings from the kelp of Neah Bay in Washington State, we reviled Andy in absentia for about the 99th time. "I wonder," I said, "how many salmon we'd have caught around kelp in the old days if we'd never listened to Andy?" S was more charitable – "I guess there were so many fish then that he didn't have to learn about kelp."...

Where Andy's misinformation originated or just when it was debunked we know not, but the revolution did come. By 1940, fishing the kelp was an established technique.

Brother S accepts the standard view that salmon like kelp because they find baitfish there. Any fisherman who avoids kelp reminds him, he says, of "the guy in the story who lost something one place and was looking for it in another 'because the light's better here.' Kelp – sure, it's where you foul your line and sometimes lose it, and where salmon foul it and you lose them, too. You have to take the chance. It's where the salmon are."

Kelp! The lessons it taught us in the lower 48 all served to prepare us for Southeast. In innumerable locations across the Panhandle, salmon and kelp come bountifully together.

"We'll never get to fish them all," laments S. "Life's too short. In all our seasons we've tried maybe one percent of what there is."...

We have names for our one percent – Vallenar Kelp, Niblack Kelp, Grindall Kelp, Island Point Kelp.... Three Kelp Hole, Big Kelp Slot, Lone Kelp Hole....

Vallenar Kelp, the bane of Vallenar Point, stands historically atop our list. Its productivity has declined in the modern day, but – what a beginning!

1953. I was alone that year, in a 16-foot open boat, piling up the sentimental riches I would use, among other purposes, to tempt Brother S to join me in the future (which in 1955 he did).

Vallenar affirmed the truth that in Southeast, man and nature face off. Within yards of me as I poked along in loneliness, a Coast Guard light blinked. Ahead, across Tongass Narrows north of Ketchikan, a building or two stood square against the forest. And beyond Point Higgins, on the way to our future campsite in Knudson Cove, Clover Pass Resort accommodated visiting sportsmen.

From a guest at the resort I got an earful about Vallenar:

"One time there I caught three salmon that weighed over 50 pounds each. What fish! But the kelp there – it's awful. And the tide keeps pushing you into it. Before you know it – foulups!"

He may well have spoken truth about the fish. He surely did of the kelp. Vallenar's kelp revels in treachery. By the fishy appearance of its visible bulbs and ribbons it invites you close, where underwater confederates lurk undulating in the eddies, rippling and waving, angling for fishing lines.

But salmon? You bet. The assortment of them in our logbooks include the giant of our Southeast careers, taken by S July 1, 1979, and weighing in at 57 pounds – 50 pounds dressed....

And Vallenar's kelp, praise it, is less bad than some. After relinquishing a few lines to it while snaking a few salmon out of it, you know – within the limits imposed by the stage and the direction of the tide, of course – where it is. Elsewhere –

Niblack. We were passing well outboard of any visible kelp bed, trolling homeward in the fishless peace of a slack day, idly watching the greenery of Cleveland Peninsula drift by. We had four lines out, two short and shallow and two longer and deeper, all hung from pole holders.

Brother S looked over the side. "Wup! kelp!" he said. "Wup! This line's hung up! Wup! So's this one! Wup! They all are!"

Backing beyond the point of contact, we wrestled the two short lines free and tackled the other two. One came loose with a rockfish on it and the other almost with a coho salmon, eight pounds perhaps. The spinner, mangled and hanging tail first on the hook, was not quite good enough for him. He made a final left-right scouting trip and disappeared....

Three Kelp Hole. Yes, only three. For density of kelp growth, this stamping grounds for salmon must rank last among all the kelp patches of Grindall – Grindall Point, Grindall Passage, and Grindall Island.

Of the countless twists and turns of shoreline here, Three Kelp Hole occupies an obscure corner at the inner end of Little Gut, a narrow pass separating a steep islet from a higher chunk of land.

We had caught fish a dozen places at Grindall, but – inexcusably in retrospect – never had tried Little Gut....

When those three kelp came abreast of our gunwale, 20 or 30 feet to port, this tiny bight of Grindall came alive. One chinook hit, and leaped, and then another hit. We landed one and lost one, circled the islet, came back through Little Gut to Three Kelp Hole, and hooked another in the same place.

Chinooks, cohos, pinks – Three Kelp Hole thereafter generously yielded them all. It became an old friend, did that beautiful, simple little piece of water with the triangle of kelp bulbs on it.

Three Kelp Hole holds one sad story...

The ribbons on the three bulbs waved toward us as, barely moving against the current, we approached. Deep water lay behind us, but here the bottom was coming up rapidly.

Abreast of the hole a 35-pounder consumed my spinner and ran for the kelp. The nearest bulb went down, came up, went down, bobbed up and down. "He's fouled," I groaned. "He's still on," S said. "The kelp's still bobbing. "Have to get over there."

With his gaff Brother S pulled up the kelp. "Why, there's a wrap clear around it!" he gasped. "How could a fish do that? ... He laid his knife against the kelp below the turn of line – and hesitated. Ruin one of those three dear kelp? But what else was there to do? He sliced. With a bubbling sound the kelp went down – and we went out through Little Gut in pursuit of the fish. An eighth of a mile away we netted him.

But Two Kelp Hole? No. It will forever be Three Kelp Hole.

Kelp! Island Point. On a break for lunch we pushed the boat onto a small, dense patch lying against a steep rock, and hung one bulb inboard to hold position. We were out of the current, and the rock and the kelp damped the swell. Relaxation. But –

"Hey!" S suddenly whispered. "There's a big fish down there!. Holy smoke, he's a 40-pound king!"

I was at the other gunwale and could see nothing. In his hushed voice S poured on the details. "He's five feet down, just sitting there, wiggling his fins.... Now he's moving... headed your way...out of sight...see him? No, now he's backed up again, right back where he was."...

I slid over to S' side and stuck my head over – barely in time to see the kelp stems flex as the fish streaked out.

An outdoor columnist of my acquaintance says king salmon "like to stay in the kelp at night, waiting to scout for baitfish at dawn." However that may be, this one was there at noon.

We swallowed the lunches, trolled back and forth in front of that mat of kelp – and caught two 20-pound-plus chinooks. "But neither of these is the one we saw," said S. "He was bigger."...

Big Kelp Slot. Our first trip through this strange passage off the west shore of Gravina Island was an adventure in discovery....

To both port and starboard, kelp stretched out by the acre. But straight ahead, a 40-to-50-foot-wide channel lay open. "It's a trap," said S. "Oh, well, let's go through."... We did, gingerly, all the way, with two lines out, never got snagged with either, and at slot's end landed a chinook.

The sessions of suspense Big Kelp Slot provided after that included a red-hot time with pink salmon.

Being salmon, pinks are great fish, but as a species they compete in fast company. Chinooks and cohos net more pounds per fish and more cash per pound. At one point in history Brother S rounded off our fish-slip prices at $50 for kings, $10 for cohos, and a single buck for pinks. "But pinks – there's a lot more of them," he said, "and sometimes you can catch them in a hurry."

We caught them in a hurry that afternoon in Big Kelp Slot. First we spotted one in the water, and then another, "and now they're showing up all around us," said S. "Looks like another run like we saw over at Bond Bay that time," He referred to a school that had gone past us – and ignored us – in about five feet of water there. Big Kelp Slot was deeper, but the pinks were close to the surface – so close that we could see the blotchy black spots on their tails.

Then the action started.

According to Brother S, "Only about one pink in 10 is of a mind to bite a herring spinner, but when they're as thick as these, one in 10 is plenty." They were four-pounders. We snubbed them stoutly, losing quite a few but bringing most aboard, often without gaff or net. They cooperated by their usual tactic of circling and bulldogging rather than running for the kelp. Before the bite ended we had landed more than 40....

Lone Kelp Hole. The single bulb that marks this spot lies a few hundred yards from Big Kelp Slot. Inside a small area here we have taken many salmon; outside it, few. It is a contemporary of Three Kelp Hole at Grindall. Says S, "There's some kind of underwater connection between these places and the main beds."

Nice theory. And he has another, for No Kelp Hole.

No Kelp Hole. We gave this snag-free rocky basin its name after a veteran Alaskan told us it once held kelp but a storm "cleaned it out." While mooching our way toward it we snagged several kelp. In it we landed two nice fish unimpeded by kelp..

S' theory: Salmon are attracted by the same bottom contours and composition as the kelp is, "so they'll be there even if the kelp isn't – we get the salmon without the kelp."

The salmon without the kelp? I thought of all the kelp beds we have wrangled with, dozens of them besides the few mentioned here. The salmon without the kelp? "What a setup," I said. "We should find lots of places like this."

"It'll never happen," S said. "Salmon without kelp? Nobody has a right to have it so good."

Things to Unlearn

Myths, misconceptions, old-wives' fishing tales. Southeast mooching will not shield you from them.

When and how do these stories get started? Does some person see some fish do some thing, assume they all do it, tell somebody so, and set the whole mill in motion?

The trouble with human beings is that they think with human brains and react with human nervous systems. Fish and wildlife use their own, far superior for their purposes. Human logic intrudes where it is foreign.

Sharks must turn on their sides to attack

I address this odd notion not because debunking it will benefit you immeasurably but because (1) it makes my point that human logic flops miserably when misapplied, and (2) it claims the distinction of having met a rare fate among fishing myths: man has actually discovered that it isn't so.

Grade school class, 1920s: The teacher, after securing our attention to sidewinding sharks by billing it as useful in avoiding being eaten by one, explained the logic with pointer and sketch at the blackboard – "His nose, you see, gets in the way of his eyes."

Delivered by unimpeachable authority, this information embedded itself in my store of knowledge and stayed there at least into teenage years. Then field experience interfered. I saw dog-fish sharks remain perfectly upright while feeding on baitfish. Further realization followed. No way could a shark so twist his body as to change his eyes-nose relationship.

In time, with the help of upright sharks in TV documentaries, the myth waned. It is dying hard. In my own little library, a fishing book published in 1974 still says that all sharks ... "must roll over almost on their backs to attack."...

Frightened fish won't bite

"When the killer whales move in," says another book, "the fishing is over."

No, sir....

The orcas were right next to my boat and seemed to be whispering to one another. I was in process of determining that the *pssstt* was actually their big dorsal fins slicing the surface – when my reel screamed salmon. I landed him and another while the whales were still in sight. I became a firm unbeliever.

Nature's creatures possess sensors much superior to ours. They react faster and finish reacting sooner. A man, on narrowly missing death in a highway crash, pulls to roadside and spends long, weak minutes wheezing his heartbeat down. A salmon flees from a marauding seal and resumes his attack on the herring.

Frightened fish? Listen to a commercial diver who harvests clams hydraulically from submerged beds: "I make a lot of noise and fuss with my nozzle, blowing junk all over – and the cutthroat trout swim through my legs and arms to eat the tidbits."

Fear in wild things must be accompanied or instantly followed by something else. I once startled a doe feeding on a Southeast beach and watched her bound wildly upland, swinging her head to snip at willow browse as she fled.

Gazelles on TV graze peacefully in front of lions, who recently ate one of them and are not hungry any more.

Some steelhead anglers toss pebbles at fish to chase them out of places casts can't reach – and then catch them. Not that this tactic always works – but surely, at times, it does. So keep fishing. Let the killer whales drive the salmon to your line.

Salmon always swallow baitfish head-first

Smart fish might have developed this idea, to get you to put your hook up front in baitfish. Human rationale says a tail-first bait would hang up on its fins in the predator's mouth. Roderick L. Haig-Brown, in his great salmon life-story *Return to the River,* says his salmon-character "Spring" seized a herring in her jaws and then "turned it and swallowed it." He does not specify headfirst, but any old head-to-tail believer will take him so to mean. And I have other literature which states headfirst outright.

Await this experience: While probing the bottom for salmon you feel the yanking of a rockfish, followed shortly by a much heavier pull; a big ling cod has grabbed the rockfish. This ling has a mean mouthful. A rockfish's dorsal-fin spines are long, sharp, backward-pointing and poisonous. A four-foot ling cod swallowing a one-foot rockfish tail first strains credulity. But by the time you get that ling to the surface he has the rockfish's tail in his gullet, and only when you have pulled him half out of water does he angrily turn loose.

At one time long ago I believed in the headfirst theory myself. Brother S never did. He called me to the cleaning rack one day to show me why: herring pointed both ways inside a salmon.

He also makes his case otherwise:

"The way some fishermen cut bait, how does a salmon know which end is which?"

It was not the only time he had asked that kind of question. One of the others was when I caught a salmon containing the herring heads and entrails he had thrown overboard while baitcutting – "which part of those things looked like the front?"

Chinooks prefer large, slow-rolling baits

"Well," says Brother S, "some of them may. What I can tell you for sure is that when I started cutting spinners instead of using whole herring, I also started catching more fish. The difference was – or at least one of the differences was – that the cut spinners spun faster.

"Almost always, faster is better, especially if you can also put in a little wiggle. ...

"The whole herring I used to use rolled over in big loops. Once I watched a salmon grabbing at one, and he never grabbed in the right place. I think he twisted his neck. ...

"A fast spin is a tight spin. Straight line. When a big fish opens his mouth he can hardly miss it."

Right-handed herring?

In the autumn of 1998 I met a fellow whose interests matched mine, and in no time we were exploring the field of fishing. He shared "some weirdness" he had heard 40 years earlier from another man and never repeated "because it sounds so far-out." But, he said, "to this day I can remember his words."

Go back 40 years and listen.

"Always fish with male herring – those with milt in them, never eggs. Nature tells salmon not to eat the females." ...

Yes, that's what he said. And he said something else:

"Be sure to cut and rig your spinners so they spin right-handed. Clockwise. Backwards ones don't catch much."...

Weirdities? Well, I took them to a research-physicist friend who, despite his advanced education, adapts well to the real world when mooching.

"Male herring only? Hmmmm. Could be... tremendous sense of smell fish have. ...

"And right-handed herring? Well, there is certainly the force of Coriolis, whereby earth's rotation deflects moving bodies to the right in the northern hemisphere. It could be that salmon prefer that direction. Or that a live crippled herring will naturally whirl that way."...

How about that! Was it the revelation? Had we caught all our fish on right-handed male baits, and never known it?

I doubt it. I'll have to test the theory, though, come summer. But I'll begin with a bias. Troll around with a left-handed female bait that's not supposed to catch anything? Doesn't sound like fun.

So just forget any idea like that, you say – they're nonsense?

Careful, now. Don't let your human logic betray you. Believe not what you think but what you observe. Discard no proposal until you have personally disproved it.

And adopt none either, however logical of sound or broad of acceptance, without seeing for yourself with your lines in the water. This trial-and-error approach offers a bonus feature. It calls for plenty of fishing.

EXTRACURRICULAR ACTIVITY

You won't hear of it elsewhere in these pages, but one year, with the help of an Alaska friend who volunteered his smokehouse, we had a smoking operation going. For this we used the smaller fish, legal to take but low-priced on the market. Here Brother S readies a batch for the smoker. To capitalize on this idea, pick a year like 1996, when a glut of pink salmon depressed the market. And pinks are delicious smoked.

8. TIME OUT!

Being a Southeast moocher, you have reason to sympathize with your underprivileged fishing friends in the lower 48. Their vacations are short and they use them catching few fish, spending lots of money, and never tiring of the game. In contrast, your vacation (that is, your fishing "work") is season-long and you spend it catching quite a few fish, sometimes making some money, and occasionally growing so weary of the grind that you take a vacation from the vacation, or at least switch emphasis. The side benefits of the life temporarily become the main course.

The Inimitable Eagle

The first eagle I ever saw take a fish out of water was so outperformed by the second one that I am still filling in the space between. What you can deduce from this is that the distractions attending your Southeast fishing will not diminish with time. Eagles will help see to that.

Perhaps, in pre-school years in your locale, you were handed down the same fables about eagles as we were, in our country circle of little kids. Eagles symbolized all that was wild and wide open. They lived far up on remote cliffs or in other mysterious high places called "aeries," and they descended on civilization only to carry away lambs and babies.... Don't believe it.

On the other hand, don't expect any two eagles to be alike. Eagles are individuals, as diverse as your human friends. You may at best be able to categorize them as to types. Many types.

Of the first fishing eagle mentioned above I wrote no praise:

"Eagles are overrated. Today I watched one overtax himself trying to fly off with a two-ounce herring. He looked great when he swooped down and closed his claws on it. But then he had to paddle along with it, until finally, barely, he gained altitude."

A couple days later I recorded the second eagle. He "took off with a rockfish that must have weighed two pounds and carried it nearly straight up to his nest 150 feet high in a tree."

The idea that eagles symbolize wildness got scrapped quickly for me in Southeast, undone by one white-headed black giant perching atop a 40-foot-tall waterfront hemlock.

I was easing up on him in my boat. I crept within 100 feet. He did not move. I was so intent on this stalk that I quit breathing. I would film him with my movie camera as he flew away.

But he did not fly. I beached the boat, walked up to the tree, and waved my arms, demanding action. He lifted one foot, swiveled it, put it back on the branch facing the other way, picked up the second foot, rotated it, and put it down, thus turning his back fully on that silly person on the beach....

Eagles exploit fishermen. "Sit on a limb, son," the parent tells the juvenile. "Sit and watch. Sooner or later the fisherman will pull in a fish he doesn't want and will throw it overboard. All you have to do is go grab it before it wakes up and goes down."

Eagles mostly debunk the sentiment that they should perch mainly in tall trees silhouetted against the sky. You will be mooching for chinooks along a meandering shoreline, thinking you are alone with the rugged world, when you feel yourself under scrutiny. Your angle of sight into the low spruce boughs changes and the big bird takes shape, watching you. His white head and tail give him away.

Some eagles know their strength and some don't. Two of them demonstrated this difference within an hour of each other one morning, upon tackling fish too heavy to lift.

The first of these swooped down, grabbed the prey, tumbled over, splashed around, and flopped, thrashing and empty-clawed, to the beach, and there hunched down in ignominy. (Maybe he was lucky to be loose. Some observers say that once an eagle gets a solid hold on prey, he cannot let go.)

The second eagle glided down, grasped his victim, and calmly towed it ashore in flight.

Some eagles go beyond that one, with a tactic we'd witnessed a few times in early years but not again until, in 1996, another one performed it.

When he foundered with the load, he improvised. He made a boat of his body and oars of his wings and he matter-of-factly rowed ashore. He clambered up on the rocks towing a 2-pound salmon and immediately began eating it, as if merely capping a routine operation. I tell you, eagles are individuals.

The assertion that eagles have telescopic vision is an understatement. An eagle can see anything you leave in the water from further away than you can see the eagle.

What you see is a speck against the sky, far away, barely visible. It grows larger. It is an eagle. It makes a beeline for the dimple on the water behind the boat where the winded rockfish you caught and released is stirring, ready to go back down. One fell swoop and the eagle is gone with it....

Eagles grow acquainted with fishermen, and perform for them. One of these – one that regularly monitored us off Caamano Point – put on a special show one breezy day. He yanked a two-pound, 15-inch black rockfish off the windblown surface and rose mightily with it at 45 degrees upward to a towering spruce top.

That was the introduction. No eagle can eat a 15-inch rockfish out of one claw while clutching a waving limb 200 feet above the rocks with the other, but this one did. A grandstander, that eagle. He knew we were watching him all the time.

Out of Adrenaline

It is not impossible for Southeast mooching to satisfy you, believe it or not, to a point of saturation.

The fix you then are in also afflicts other professionals, surgeons for example. You may have met, or even have been served, by one who illustrates the point.

He works 12- to 16-hour days. But he also takes vacations to get into something like mountain-climbing or sailing, hauling on the halyards and heaving to the seas to rest up.

While surgeons are not to be compared for skill with moochers, the principle applies. When you are commercially licensed and therefore working at fishing, fishing may begin to seem like work.

ODD FIND: Trips ashore may yield special discoveries, such as the totem pole Brother S has come upon here.

This is when, somewhere out there, you find a place to go ashore and, as soon as the land quits its sea-induced swaying beneath your feet, poke around awhile where no one lives, or but a handful do, or once did and no longer do, as at abandoned cannery sites.

"Moonlighting," Brother S calls this adding of a second recreational activity to a first.

You must keep your moonlighting in perspective, lest you sacrifice the ability you once had to recall every chinook over 40 pounds and every hot bite of cohos. Memories of events ashore will try to crowd them out. Among the delights we've catalogued are these:

– The cave in the cliff above the high tide line, dry and calm on a rainswept day, where the appetites of fishermen made the can of beans a banquet...

– The pile of mussel shells on the dirt floor in front of the rusty stove in the brush-grown old shack – the evidence, Brother S deduced, of some soul marooned and "eating mussels on the half shell in a storm that must have blown a week."

– Another remote cabin, this one with a "welcome" mat outside and, on the walls inside, framed scenes of retreats in wildernesses ... Brother S stepping outside, taking in the scenery, and remarking, "Beautiful pictures they've put up in there. But I don't know why they need them when they've got this."

– The harvests of wild asparagus, delicious boiled, gathered from a spit at salt water, and of Hudson's Bay tea leaves from a muskeg flat, where young frogs sprang up absolutely everywhere, dozens per square yard ... a tenth as many as the mosquitoes

– And red huckleberries, releasing their tart juices inside the breakfast hotcakes.

– And wild strawberries from a sandy plain beneath the forest and the hills, much more tasty than any domestic model from the supermarket but so tiny you couldn't get enough.

– The broad and solid carpet of blue lupine, lush and beautiful, spreading inland behind the drift logs, but a sight never to be seen unless you went there.

– The hot tubs awaiting weary boatmen in the bathhouse in Warm Springs Bay ... and the discovery that the pipeline feeding them had changed the climate of an infinitesimal part of Southeast: where a leak in the line sprayed steaming water into the forest, tropical foliage was growing....

– And one more old cabin, hidden ingeniously in the forest. We learned it was there only because we were told. Although we can see the water from it, never have we been able to spot it from the water....

– Finally, among pictures coming to mind at the moment, a squirrel perched at eye level on a huge fallen tree, declining to retreat when we approached. We made fast motions at him but he stayed put, stiff-legged. Confrontation. The little animal won. We backed off....

You can't reap rewards like these without investment. Like the surgeon who must go to sea, you must go ashore.

Your findings can be as monumental as your attitude enables. Take the time when, on one small island among many, we beached the boat at a tiny landing place available only because the tide was down.

A cliff rose nearby and we climbed it through thick brush from the side. The summit gave us a park-like view. With a railing to prevent the 80-foot fall? Hardly. Brother S stepped solemnly to the edge, taking in the panorama.

"We have just made history," he proclaimed. "No human being, I'll wager, ever set foot here before."

No one, ever? So what? In every direction lay spots where no one had likely ever set foot. But again, so what? If you're enjoying yourself and you want to make a monumental event of a little one – why not?

Uninvited Guests

What a moocher seeks from the sea is salmon. What the sea presents may be salmon or an array of other creatures. Surprises follow surprises. One morning Brother S lost an outfit to a sea lion. The same afternoon I lost one to a harbor porpoise. We don't know whether the sea lion struck or got snagged, but the porpoise bit on a small cut spinner.

On the line he swam around porpoising. He wouldn't come in for unhooking, and my 20-pound line parted.

That day was not a phenomenal one, in the total scope of Southeast marine wildlife. Here are a few other samples:

ROCKFISH. These quill-backed customers come in such assortment you'll never be able to differentiate all the sizes, shapes, and colors. Big ones, medium ones, small ones. Some thin and deep, some husky and heavy, many in between. Some uniformly colored and some atrociously mottled. Black rockfish, red rockfish, banded rockfish, about 40 species in all. There will be few days when you don't encounter some or others of them. Your main concern – unless you decide to become a serious rockfish fisherman – will be to get them safely turned loose.

DOGFISH. These small sharks (small? up to five feet long) are great troublemakers. Since you are fishing with natural bait, and especially since you are doing it at slow mooching speed, dogfish achieve their enthusiastic worst. Not content to rob your bait, they also snip your leader.

Keeping in mind that dogfish-attracting baitfish also draw salmon – how many leaders are you willing to sacrifice to get through to the salmon? Decisions, decisions!

FLATFISH. Among these bottom dwellers, meet the soles, sand dabs, flounders, and turbot, plus the granddaddy of all, the Pacific halibut. Four-hundred-pounders exist.

Don't expect to land one this big on sports gear, but you might pull up one of 100 or so. Pull, indeed. Usually he comes up flat side first, for best resistance, so your pulling gains only a few feet per minute. Then he swims back down and you pull again.

An occasional giant opts for all-out horizontal travel, unrestrained by lightweight monofilament nylon. Certainly many of the "monster chinooks" which sports fishermen cry "ran my line to the end and then broke it off" were in fact monster halibut.

LING COD. One scenario involving these brown-mottled rocky-bottom dwellers will repeat itself often enough so that, after being mystified the first time or two, you will feel-read the script right through your fishing line. The example we relate here we choose because the ling was our largest.

..."Wup!" said S. "Got another rockfish on. Hit bottom and didn't pull clear fast enough....Now have to get him up and take him off. ... Wup! Too slow again. Little fish got grabbed by big fish. Ling cod."

After several short runs and a few bulldogging sessions, the ling let himself be pulled up. He was four and a half feet long and over 70 pounds, a Neanderthal customer with a head that would hold yours and fangs an inch long. He was not hooked, simply hanging on. He gave us the evil eye, let go, and sank away.

SABLEFISH (BLACK COD). You may go for years without ever taking one of these. Then a season will arrive when they are everywhere. We saw our greatest concentration in 1978.

They were foot-long, green-hued juveniles. Over an area at least 10 miles wide and 20 long and we know not how far beyond, no place was free of them. They were voracious. We hooked them at every cast, often two at once on our two-hook rigs. "No wonder they're hungry," said S. "There's no room between them for anything to eat."

The good news was that, perhaps because of the competition, other fish were hungry too, and when, close to the boat, we were able to get our lines down, we caught salmon. The year turned out to be one of our best.

MAMMALS. They can come in bunches.

Calm water extended beyond Chasina Point – a treat in itself. The point is open to Clarence Strait, and therefore to Dixon Entrance and the Pacific. Scant stirring of the air is required to rile the waters.

The forested shores of Cholmondeley Sound retreated behind us as we worked eastward along the north shore of the point.

In the strait ahead a whale spouted, and another. Humpbacks. We see them occasionally hereabouts, and on one occasion gasped at the sight of two of these 40-foot behemoths leaping completely clear of the surface and spreading water a city block when they crashed back down. (Such antics may be a "whale to whale communication and a show of strength," according to January 1979 *National Geographic* magazine.)

We were treated to no aerial show this time. The humpbacks spouted on toward Dixon Entrance. But now came killer whales. They appeared suddenly, one and then several, giving us the eye, curving around us like sharks in a horror movie. One fin looked six feet high. It came within 40 feet, slicing a vee in the water with a crisp hissing noise – "*plpshhhhh!*"

"I don't like this," I said. "They'll go away," S said. "*Plpshhhhh,*" the fin said.

They did go away. But when there is nothing but clear space between one of them and you, it is hard to feel sure he will.

"I admit it was a little scary," S said...."Humpbacks and killers the same afternoon. Now all we need is some porpoises."

WHALE SHOW:

He heads skyward...

Clears the surface...

And crashes back down with a splash that spreads water for a city block.

No porpoises showed up that day, but just to avoid being totally uncooperative, they did put on a show within the week.

It was mid-afternoon. We were dead in the water, doing some up-and-down trolling, when they appeared, eight or 10 of them.

You will often find porpoises pacing your boat while you're traveling. But in a whole season you may never watch them do what these did around our motionless craft. For 20 minutes they swam, splashed, and cavorted around us, showing their black-white hides like paint horses, telling us we know not what.

One of them, broaching within 15 feet, lifted my line out of water as he swam away.

"Trouble!" said S. "Just wait 'til he hits the hook!"

The wait lasted half a second. The reel snorted once, but that was all; he didn't get hooked.

Then, without explanation, they all disappeared.

Ashore that night, veteran handtroller Bob Stanker told a tale of porpoises just like ours in another location. "You should have seen them. They were circling my boat, rushing and splashing around, a lot wilder than they usually are. Don't know what they're thinking when they do that."

Good. Neither do we.

Stormbound

Until the low barometric pressures brought the high winds down Clarence Strait and drove us and our 20-footer to cover, we had been profiting substantially on chinooks. Big chinooks. The late-season ones of July.

"The run was petering out anyway," Brother S said as we fled the building seas. "We got in on most of it. Can't complain."

This gathering Southeaster was no surprise. Warnings had been broadcast from the NOAA weather station on Annette Island, and had helped us choose Island Point for this day's work.

Island Point extends into Clarence Strait from the east flank of Southeast's biggest island, Prince of Wales, and is named for a little island marking its end. Its most productive mooching stretch is short – a limitation all to the good, this time. At worst we would be three minutes from the safe lee side of the point; at best, 30 seconds. We could haul gear first if we had time, and afterwards if we didn't....

The clouds turned the color of a tar-fire smoke. The air thickened. The puffs of the coming Southeaster smote us. That was when S said, "We gotta go," and we hauled gear and went.

Two miles down the sheltered shore we turned into the narrow pass that enters Trollers Cove. Behind us in the strait by now, whitecaps were being turned to spray and hurled northward over open water. We picked our way through the pass, scouted around in the sanctuary inside, picked a deep-water spot, and dropped anchor. We would be here awhile.

I might theorize that Brother S' "can't complain" philosophy, mentioned above, helped condition us to accept the Southeaster's usurping of fishing time and encouraged the other use to which we put it. On the other hand, no one really knows why certain moods arise. But when one does, I tell you, make the most of it.

Peace fell on the world. Quiet peace. Out along the strait whence we had come, we knew the spruces were whipping and waving, straining at their rocky roots, bending, bowing, growling, groaning, snarling.

In Trollers Cove we sat in security in the warm galley of our craft, two gray-headed old geezers, facing each other from opposite sides of the miniature table, knees angled for room and faces two feet apart. Silence.

Silence? The wind roared. But the noise was far away and high up, all around but not close, a contrast to the stillness of the inside. The spell increased.

Childhood on the farm. Parlor fun. "Can you still wiggle your ears?" I asked. The ears went back and forth. "And your scalp?" The gray hair swung forward and back. "Never did understand it," I said. Neither achievement had ever been mine.

"But I bet you still can't roll your tongue," I said, and demonstrated with my own, making it a protruding tube. His still came out flat – nothing he could do about it....

No two Southeasters are quite alike. When the subject matter is serious you may even figure out ways to save the planet. But this weird session was more fun.

History in a Hurry

In the fortress of jagged black rock guarding the forest from the sea you detect a crack which, on approach, becomes a passageway, and you ease in. The sidewalls spread away and you land on a pebble beach fronting a miniature meadow. Southeasters could not reach you here. You are in one of those little worlds known as "God's Pockets."

While the boat that links you with the other world rests its bow on gravel and waits patiently, you nose about. The layout makes perfect sense topographically. A gully down the hillside ends here; the meadow is a delta. Of course it is. The stream that carried the silt that built it still gurgles forth from the moss-hung spruces.

Shelter, water, flat ground – completeness. *Great place to build a cabin*, you think.

The toe of your boot nuzzles at a flat patch of lichen and – what's this? You scrape more moss away. What indeed? The collapsed remains of an old shake roof. Build a cabin? Good idea. But at least half a century late....

What fisherman, hunter, trapper or prospector once squatted here? You've little chance of learning, since, as Brother S says, "there is no courthouse file."

Brother S further notes that in any case we two johnny-come-lately moochers can have little first-hand knowledge of Southeast's past. But the picture seems thus: Fisheries in Alaska, like farms elsewhere in America, were once numerous and scattered: fisheries for which fishermen set up camps, and companies built canneries, where the fish were.

Since then, improvements in boats, gear, and communications have cut the number and boosted the size of the plants, leaving the old ones to go back to nature.

Similar fate has befallen various cabins and shacks that await a curious eye here and there. Weathered and tumbling, they are in process of disappearing.

"But until they're gone we'll keep finding them, said S. "Can't miss them. They're in the only places where we – or whoever built them – could get ashore through the rocks."

So while your flattened shake roof has been fun to uncover and while you've enjoyed a break from workaday mooching, you have made no great archeological find.

But you have sampled Southeast history, recorded or unrecorded: how much of it has been made, by so few people, in so little time....

FUR FARM
NO TRESPASSING

Signs like the one above used to greet us from certain small islands as we mooched along. They dated, an old Alaskan told me, to Prohibition days.

Some of the signs, faded but legible, were still in evidence at least into the 1960s. Possibly, if you went ashore and searched, you could find remnants today.

"The idea," my informant told me, "was that on an island the fur-bearing animals couldn't get away. – no fences were necessary. But there was some skullduggery. Fishermen got just as thirsty in those days as they do now. And some of what the fur farmers were producing behind those Keep Out signs – well, it wasn't fur.

"But then, in 1933, Prohibition got repealed and the fur farm business was ruined."

In another casualty of changing times, we moochers lost a great fishery with the outlawing of fish traps in the late 1950s.

These money-making devices of floating logs and underhung chicken wire were positioned where the fish ran.

Where could you find better trolling than along the submerged wire fence that led the fish to the pot? "Fishing the trap lead" was a secret of success.

The traps gave us another advantage. The season for them was much shorter than for us, and the traps when idle were towed into protected coves for storage. At night we tied up to the inshore side, the securest of moorage.

Atop the logs, the watchman's quarters or "trap shack," complete with bedsprings, invited an overnight stay.

When all but Native-owned fish traps were outlawed following statehood in 1959, reminders remained. On the shore of Bond Bay on Cleveland Peninsula some of the multi-ton cup-type trap anchors had been unloaded, their rusty shanks reaching 10 feet skyward and rusty chains draped down.

In the 1970s the anchors disappeared. Where they went we never learned. We simply wrote them off to a history still relentlessly advancing....

Other changes, in comparatively modern times, have come about via a U. S. Forest Service program of building cabins for public use in remote nooks where no evidence of human encroachment previously existed.

Among these places is Trollers Cove, a beautiful retreat snuggled into the east shore of Prince of Wales Island. It has been referred to previously herein and will get more space on a future page. Trollers Cove is a great place. It was made so by original creation and remains so as this 20th century winds down. But it is not quite what it originally was.

CASUALTIES: Here are sites of two Alaska salmon canneries lost to history. The wharf in ruins is in Gambier Bay and the row of weed-grown cabins at Tyee, both on Admiralty Island. To emphasize the passage of time, these pictures are by no means recent ones. They were taken back in 1957.

OUT OF BUSINESS: Brother E stands among trap anchors unloaded ashore, "rusty shanks reaching 10 feet skyward and rusty chains draped down."

CONTRAST: As the above cabin and outbuilding at Trollers Cove indicate, the place is not as unspoiled by civilization as once it was. But as the view below from the porch shows, it is still a pretty nice place.

9. LAUGHING AND CRYING

Variety is the spice of life, right? Right. Especially the life of a moocher in Southeast. The joys, jackpots, woes, messes, euphorias, ecstasies, and psychoses that punctuate this life would fill a book – or at least the upcoming part of this book.

Professional Incompetence

Brother S trudged up to our tent one night to find a stranger boiling potatoes inside.

It will take me a minute to progress from S' shock then to the revelations upcoming now, but be patient. In 60 seconds of reading you will be spanning 50 years of linkage....

Elsewhere herein I imply proficiency at mooching, and from my half-century of participation, beginning in the infancy of the art, you could presume it anyway.

I therefore admit to many scenarios assuring my humility. For your sake and mine, I here review only three.

Scenario 1

Return with me to the 1940s on Puget Sound.. The state-of-the-art mooching bait at the time was a beheaded or "plug-cut" herring (still no slouch today) and I aimed to become a master cutter of these. With sharp knife and calculating eye I tried all angles, seeking perfect action.

I think I've got it, I said to myself as I watched one of my creations roll neatly in test alongside the boat. This irresistible bait I then deployed. In good salmon water it went down, came back up, went down again, and again and again, until it was uselessly limp. Then a 15-pound chinook ate it. And there's more. Out of that salmon at the cleaning rack came all the herring heads and entrails I had thrown overboard while cutting my perfect bait.

Scenario 2

I persevered and, despite setbacks like the one above, had some acceptable pre-Alaska years, until in February 1951 I confidently expounded on mooching in *Field and Stream* magazine.

The deflations that followed were capped several years later by a man fishing near me off The Pup Island in Southeast.

He was in a small boat rented from Clover Pass Resort, and he sported a new rod and reel. I had never seen him before.

I hadn't caught anything. He boated a big chinook and proudly waved at me. I waved back.

He landed another, then swung close to me.

"What sport!" he exclaimed. "Wow! I'm from Stockton, California. Never fished for salmon before. Wow! Have you?"

"A little," I said. "Your system sure is working."

"I found out just how to do it in an old *Field and Stream* magazine," he said. "Some guy named Jensen wrote about it."

I didn't let him in on the clincher.

Scenario 3

The man Brother S found boiling the potatoes was one Don Cooper, burly 6-foot, 4-inch logger out of DeBorgia, Montana.

You may since have met Don in your local opera house or auditorium. For more than 30 years, on into these 1990s, he has been starring on the national travelogue circuit with movies featuring the adventures and misfortunes of himself and his brother Dennis in wild Alaska and elsewhere. He turns lessons in the history and geography of a place into hilarious entertainment.

Slapsticks are not beyond him. His audience, skillfully preprimed, howls with delight when he falls overboard from a boat. He is a genius at acting inept and ignorant about everything.

Cooper and I first met in Juneau in 1953. We fraternized during the few days he was there, and afterwards maintained contact from widespread places. Plans resulted. I visited his home under the Big Sky in DeBorgia – and he boiled potatoes in Brother S' and my tent in Knudson Cove, Alaska. He was still more logger than moviemaker at the time, though already mastering the art of ineptness and sensing its box-office potential.

He was also supposed to be in Guatemala, South America, that week. Being 5500 miles off course was easy for him, but I had neglected to tell Brother S that he might show up some time.

"Hello-o-o!"... said S, upon lifting the tent flap.

"Hello," said Cooper. "I thought this was Jensen's tent."

"I'm Jensen."

"You've changed some."

I have never heard Cooper's version of this encounter but it must be a dilly. He sees invisible details. I have never caught him in unmitigated falsehood, but – how he can customize the truth!

While Brother S and Cooper were trying to organize the facts, I arrived carrying a 51-pound salmon.

"This is Jensen," said Cooper. "Look at that fish! I never thought he really caught fish like that."

Cooper stayed a week shooting movies of salmon on the line, deer on the beaches, and boats in front of mountains. And with Cooper's camera I shot Cooper flipping a hotcake. The cake missed the skillet and draped itself juicily over the coffeepot.

"What a mess!" said Cooper. "What a mess. Waste of film.... You got the whole thing, didn't you?"

Before departing, Cooper decided he, too, must fish. Here, Don's ignorance act was no shade short of perfect. Or else it was no act. All he claimed to know about mooching was one word, and that one was wrong.

Don Cooper

For reasons unknown to me, some fishermen call a plug-cut herring a "choked" herring.

"A choker," said Cooper, is a cable you hook to a log to yard it out. Why chokered?"

"Not chokered, I said. "Choked. It means plug-cut. I don't know why. Here. Let me cut one for you."

"I'll cut it myself," he said. He was wearing mittens. He hacked a herring in two, hung the aft half on the hooks, threw it over, and watched it drag along on its side.

"That looks awful," I said.

"Looks all right to me," he said. "I'd eat it if I was a fish."

"It's supposed to roll over," I said.

He grunted and lowered it out of sight.

"That's the worst bait I ever did see," I said.

"Something has grabbed it," he said. "Here. Take this and play him. I gotta get some pictures."

Thirty-six pounds.

His next ugly bait got to deep water before the rod bent. It stayed bent. "Bottom," I said. "Bottom's moving," he said.

Forty pounds.

Plenty of pictures, sandwiched into scenes shown nationwide.

Big deal for him. And one more lesson in humility for me.

Gear Fights

September 1981. Nightfall. Roger, a handtrolling acquaintance, guided his gurdy-equipped 19-footer into his berth near ours and unloaded his fish. He had 18 cohos, a modest catch.

"I pray the run stays weak," he said. "I need the money."

He couldn't have said that? Yes, he could. Let me tell you about that tradition of Alaska commercial fishing, the Gear Fight.

A Gear Fight rages whenever a fisherman of any gear type – seiner, gillnetter, troller – perceives that any fisheries regulation hurts him and helps the others. In other words, most of the time.

Directly or in attitude engendered, Gear Fights affect every fisherman. Physical combat is not necessarily involved, but fighting spirit is.

Brother S nurses a personal Gear Fight with purse seiners.

"They're nice guys until they're seining." he says. "Then they think they own everything. To them, handtrollers don't count. Get out of their way or they'll seine you in."

Gear Fights first came to my attention when I began newspapering in Alaska's capital city of Juneau in 1952, but they were already old stuff to old-timers and an old headache to the U.S. Fish and Wildlife Service, then the regulating agency.

"Gear Fights," an FWS official told me, "have been around since at least 1920, back when everybody but fish trap owners thought fish traps were getting more than their share of the fish. Gear Fights now are more mixed up. Harder to handle."

When Alaska became a state in 1959, Gear Fights came with the territory. Some aspects of Gear Fights remain unaltered into these 1990s. One is the sure knowledge of every gear-type of fisherman that he is the good guy and all the others bad.

All gear types, of course, traditionally heap criticism on the state and its fisheries regulators. Juneau as the state capital gets the flak as if it were a person – "Juneau doesn't have a clue about what should be done"; "Juneau is messing us up again"; "Juneau is a joke."

Juneau is a genius, to cope with all the guff. Gear Fight – Juneau itself coined the term, an FWS official told me in private in 1952. Salvation! Without it, how could anybody get through the job of deciding which gear type could fish where and when, and stay sane? With it – a degree of relief. Let the seiners scream of a Juneau-gillnetter conspiracy. Juneau could at least comfort itself with shrewd analysis: "It's just another Gear Fight."

We handtrollers have known bliss. For decades, while Gear Fights raged, we smiled at the turmoil. Except for some close-in hotspots where organized sports fishermen won exclusive rights, we roamed unfettered.

But as equipment grew deadlier, fishermen smarter, runs thinner, and Juneau sterner, the stuff of Gear Fights trickled down to us, and in the regulations for 1979 the gavel fell. Handtrolling would be suspended for several 8-day periods.

Good heavens, Juneau! While power trollers combed the water with four lines weighted with 50-pound balls and spangled with clattering flashers, we with our flimsy lines and single spinners sat and watched, regulated out.

The 8-day provision was junked in a following year. But another replaced it. We were subject to day-by-day regulatory adjustments based on field checks and promulgated by radio.

One broadcast overtook us while we were being deluged by hordes of ravenous cohos. We had put more than 30 aboard and I was fast to another, oblivious to the world and the radio when time for the daily word arrived. I landed the fish presently.

"You missed the broadcast," said S.

"Too busy. What did he say?"

"He said there are no fish. They're closing us down."...

What of Roger's prayer in 1981 that the run he'd got into would not increase? He reasoned that if it did, "the purse seiners will be turned loose and there'll be no fish left for us."

(The prayer was answered. The run stayed lean, the seiners were restrained, and Roger prospered).

As for the idled seiners – did sympathy for them defuse Brother S' Gear Fight with them?

Hardly. There can be no truce until they make some effort to be good guys, he says. Meanwhile one of the milder of his commentaries on them goes about like this:

> When a purse seiner decides to travel somewhere, the sole requirement for the course he sets is that it be occupied by a handtroller. Collision does not result. The near miss is what titillates him. The only design specification a purse seiner demands in his boat is that it make waves like a battlecruiser, only steeper. So you can't retaliate. Stand up to pitch your anchor through his windshield and you will be pitched overboard by the wake. You can't even protest to Juneau. Arbiters there will quash your case summarily – "It's only another Gear Fight."

The Self-bailer

Near the middle of the next century, when you are as old as I already am, and are looking back on the good old days of now, tell me – why were they so good?

No, let me tell you. Because they were not so good. The logistics of a Southeast mooching season were formidable. You appreciated what you achieved.

Ultimately you will confirm this principle:

Early experiences, whether illustrious, disastrous, or hilarious, are the ones that embed themselves in memory. Those that follow cannot dislodge them. The originals are the immortals, be they yours in the 2000s or mine in the 1950s.

From mine, take one weird, wonderful day – or rather a night and day – in 1953. Brother S at the time had not yet fled the Lower 48, and I was mooching the waters of Southeast alone. My boat was a 16-foot plywood Norseman runabout on which I erected a pup tent at night.

Pup tent on a boat? Sounds like an invention of Don Cooper, the character/logger/moviemaker/lecturer/joker of a few pages back, who once told an audience he wasn't born in a barn "but we moved into one as soon as we could afford it." After I had purchased the Norseman and a big motor and a little one, the pup tent was the only shelter I could afford. To close its entrance I hung a drape.

SHELTER DELUXE: .."And in the deluge, not one dribble did this pup tent leak."

What a delight, what a peak of economic practicality, that pup tent! With the framework I devised I could pitch it in 60 seconds of an evening and collapse it in 30 in the morning. Free fishing room then surrounded me, 360 degrees of it.

I cooked on a gasoline stove (not with Coast Guard approval) and slept in luxurious comfort on an air mattress in a bunk.

Came that night.

I was anchored snug in forest-rimmed Trollers Cove on Prince of Wales Island. The heavens poured and poured and poured, and not one dribble did my pup tent leak.

Now, if you are among those of us who squirm with delight when rain pelts the roof of a standard domicile, you will shine, glow, and tingle under a pup tent on a runabout. Wonderful, wonderful. Wonderful right into the sleepy morning when, in the four inches of icy rainwater my fun-loving Norseman had fed to its covered front from its uncovered back, I deposited my bare feet. Wasn't sleepy any more. Laughed aloud in delight, and laughed again at the sound of the laughter. Euphoria!

And now I discovered a problem with my larder, awash in a box in the bilge. "All my cans are losing their labels," I said gleefully. "Whee! I'll never know what I'm going to eat until I open it." And I rocked the boat and watched the labels wave.

Chili for breakfast? But – hey! – time to go fishing. Let the labels do whatever they will. Go fishing. Meanwhile get the water out. Use the self-bailer.

A self-bailer is a hole at the bottom of the transom from which you remove a plug and drain the water by riding high at planing speed. Some young boatmen may not know about self-bailers, but most oldtimers do. Most also have slowed down to start fishing without remembering to replace the plug....

After running a crooked course to make the draining process end where the productive salmon belt of Island Point began, I sunk my gear. No nibbles. All I got was another flood, three inches before I noticed.

Amusement no end! I could self-bail again, skating on into useless territory a half-mile south. This time I put in the plug. I also put out my gear.

The climax I thus triggered did not need its first-year status to be unforgettable. From that fishless water I mooched up three chinooks averaging 35 pounds apiece.

Misery to Spare

When you say anything in a newspaper column you risk retaliatory letters to the editor.

The letter to which I now refer was received by my editor in Bremerton, Washington, and relayed to me at our fishing camp in Knudson Cove, Alaska. This letter writer thought I spent too much time on "fun," at the expense of "the problems they are bound to get into on a trip like that." So I corrected the oversight:

> A period of hot fishing was upon us but so was a balky trolling motor. We hung it on a plank spiked to a piling and angrily went to work on it.
>
> Two rainless days had hatched a new generation of insects, and these now descended upon us. We flailed at them with rusty pliers and greasy paws and got steadily madder.
>
> "And to think that all winter we look forward to this," moaned S. "All we need now is for it to start raining again."
>
> Two minutes later it was raining.

MORE GRIEF: Brother S battles one of the balky outboard motors.

Actually we don't much mind rain. But readers are important. So I provided them further fare:

> If you've never tangled with Alaska's insect armadas, you don't know what you've missed. Fat ones, thin ones, stringy ones, striped ones, big ones, little ones, bold ones, sneaky ones, buzzing ones, singing ones, silent ones, hairy, shiny, sticky, itchy and greasy ones. Climaxed by no-see-ums.
>
> A no-see-um is a vicious carnivore of quite small size. You can see a no-see-um all right, but you rarely do, until after you feel him. He is that little black flyspeck in the middle of the pain. With greasy paw you smash him into the hole he has dug, but by then you have the pain two other places.

> "Ouch!" cries the no-see-um victim. "What do these things eat when I ain't here?"...
>
> Who do no-see-ums and mosquitoes think they are, anyway? Humans are earth's designated dominant species, aren't they? A 1000-pound brown bear admits we are, usually. Why not a 1/1000th-gram no-see-um or mosquito? When you swing at him he ought to understand he should go away. He doesn't. He has no respect for you. None for 1000-pound bears, either, I'll bet.

But back up. Before undergoing Alaska hardships you must endure getting to Alaska. You and all your stuff. And before that again, getting ready to go.

If you are good at logistics and oversupplied with patience, you can somewhat reduce the trauma of this phase, saving strength for what's to come.

For us, what's to come is 1000 miles across British Columbia in a pickup truck towing a 20-foot boat, or two pickups towing two boats, per the plans for the season: 500-miles-plus (Route 97) north to Prince George and most of another 500 (Route 16) west to Prince Rupert, where we put the rigs on an Alaska Marine Highways ferry for a 6-hour, 90-mile voyage to Ketchikan.

The trip yields fewer memories than it used to.

The first time we made it, in May 1955, crossing Canada took four days, towing no boats. From Prince George 150 miles west to Burns Lake, what had been blacktop was mostly frostboils and spring-thaw muck.

For the last 300 miles the road was almost all dirt. A little of it was two-way but this part was occasionally occupied by cattle who could not decide which half was ours. The rest was one-way.

And delightful. Tight against the deciduous forest we pressed when, a few times a day, we met another vehicle.

At Prince Rupert we parked our car and, with 300 pounds of excess baggage topped by two 15-horsepower outboard motors, hopped to Ketchikan on an Alaska Coastal Airlines Grumman Goose. We clamped the motors on rental boats, ran 12 miles north to Knudson Cove, pitched our 9 x 9 umbrella tent among the muskeg-grown skunk cabbages and huckleberries, and were in business.

COMPETITION: Between Burns Lake and Prince Rupert, some of the dirt road was two-way, but the cattle using it could not decide which half was ours.

The lovely old dirt road to Prince Rupert is since gone, replaced by pavement. That's one problem. Canadians drive like crazy. One trip, when Brother S' boat slowed him to 35 mph (56 kmh) on a hill, a Mounty pulled him over – "You're holding up traffic."

We were more than willing to go faster when we could; anxious, in fact. That's always a problem. Get there and get to fishing. Drive from daylight to dark. The reservation on the Rupert-to-Ketchikan ferry (the Goose won't haul pickups and boats) must not be missed, so you drive doggedly to get there ahead of time. Then you try to sleep while you wait. Can't be done. You are already rumdum and now get more so. Meanwhile, interruption of routine has thrown your digestive system all out of whack.

By the time six days or so later when your metabolism has adjusted, other troubles bloom. If you are wearing prickly woolen longjohns, the prickles apply themselves foremostly to the bodily places most tenderized by civilization's comforts.

This is not the worst of it. Neither are the granules sticking on your eyelids or the crusts immobilizing your lips. Jehosephat, how painful city-softened hands can become when subjected to all-day, day-after-day exposure to chilled Southeast salmon water and related abuse.

With each day they get more beat up and burned up – wind-burned, rope-burned, fish-burned, rain-burned, gasoline-burned, sparkplug-burned, sunburned, and salt-burned. They are cracked, bruised, battered, hook-punctured, and blood-blistered. From your split fingernails to your knobby wristbones and back again to your flapping hangnails, they cry for kindness. Each morning when you contemplate moving these C-clamps on the ends of your arms, you know it will be impossible.

The human body is wonderfully and fearfully made, and adaptation and recovery eventually will prevail. It takes about two weeks, or three if you're over 60. Recovery is not really the word; restructuring comes closer. The soft, citified palms and fingers are now knobby, gnarled, black-callused ugly mitts.

"I'm lucky I'm not a brain surgeon," I told Brother S.

"Or your patients are," he said.

In a standard season your personal problems are equaled or exceeded by those of your equipment, like outboard motors.

Our knowledge of the failings of these began accumulating with one owned and honored by an older friend, acquired by him in the mid-1920s and built sometime before that. It had an exposed flywheel on which was mounted a knob for cranking. It had a reverse, too; ran in the direction cranked. It was called a Motorgo. Our friend's affection for this motor was not shared by his father, who called it "the Motorstop."

Outboard motors have not stopped stopping since. Or declining to start.

A brand new one with which we replaced one of our non-starting ones while trying to save the peak of a chinook season would not start either. In the works to which the starting rope was attached were some ratcheting little steel balls which so ratcheted that the rope could not be pulled. The dealer took care of the balls but by then the fish run was petering out.

Once when hard up for a trolling motor we bought a "tested and guaranteed" used one and it ran fine, except for being permanently locked in reverse gear.

One big motor broke a crankshaft and another a con rod. A third blew its power head. Still another separated itself from its clamps on the transom. It jumped off and sank sizzling behind us. We still have the clamps – they're good for holding boards together while the glue sets.

Years ago Brother S gave names to our engines. The running

motor became "Ol' Big" and the trolling motor "Ol' Small," terms later condensed to plain Big and Small. One year Big was a 55-horse four-cycle job and Small a six-horse two-cycle.

Big made fast friends of us at first. He took straight gas in the gas tank and straight oil in the crankcase. Convenient. He would push our 20-footer at 20 knots on four gallons an hour and troll on a pint an hour. Pleased with his performance, we seldom used Small.

Big came equipped with an alarm system. If his oil pressure went down or his temperature up, he blew a whistle and turned on a red light.

We were nearing Knudson Cove with a load of fish when – "*Wheeeeeee*!" Big whistled.

We stopped and checked. He had oil and he didn't feel hot.

"Still... well, I guess we better start Small," said S.

Since Small had done no trolling for awhile, he was anxious to do some. Trolling only. Faster than that, he quit.

"High speed jet's plugged," I said. "Hand me that screw-driver. I'll see if I can find it."

To my amazement, the plug which I took to be the access to the jet was in one of the few places on any piece of machinery manufactured in the previous 30 years where you could get at it.

An advantage? Hah! The plug, once unscrewed, sprang out through the clear space, followed by a spring and something else. All these parts plunked into Clover Passage like rain.

"We've got an oar," said S. "We can paddle in."

"If we do that," I said, "somebody will come along to tow us and we'll have to explain how we got into this fix."

"Try Big again then," he said.

We tried him and – "*Wheeeeee*!" he whistled. We got stubborn and drove him ashore whistling.

We loaded him in the camper aboard the pickup, tied him down, and headed for a Ketchikan repair shop. On a bump he broke his moorings, kicked up his skeg, poked out the rear window, and dumped his oil on the camper carpet.

"Nothing wrong with the cooling system or the oil pressure, the repairman said three lost days later. "Whistle shorted. Whistling when wasn't supposed to."

No end of problems. Outboard motors. No-see-ums. Immobilized hands. What else? Stuck reelspools, stripped levelwinds.

Fouled lines – coated with seaweed, hung up on kelp, hung up on bottom, hung up on the boat, tangled up by currents. Dogfish, lost fish, poison jellyfish. Wind, cold, rain, busted windshield wipers. Problems. In climactic end to one exhausting day in June Brother S reached for the cooking oil, missed, and fried a can of chicken in liquid soap. I've barely started but I can't stand to continue.

Itching and Scratching

What makes spring so great in Alaska? Winter, of course. Winters like the 1952-'53 one I spent as a newsman-fisherman in Juneau. Too bad Brother S wasn't there.

For weeks the temperature lurked around zero. In the heavy air of that marine clime, it was very, very cold. For chill-factor emphasis, the "Taku wind" blew (this wind screams from the north across Taku glacier into Juneau). At Douglas, across Gastineau Channel from Juneau, wind and tide buried the shoreline in riprap ice.

When at last spring came and the people resumed regular outdoor pursuits, they did so with pent-up zeal.

Their reaction, you'll assure me, was perfectly natural. Brother S so appraised it, too, when I mentioned it to him.

"Anytime you can't do something for awhile, you like it better when you can," he said.

He added, "Say you are on a highway heading fast for home – have to go to the head – and you get stuck in traffic. But at last you make it – Whew, what a pleasure!"

"I've been there," I said.

"To enjoy relief you have to suffer," he said.

He would have loved my Juneau winter. But he says you can get the same sort of benefit voluntarily. As a boy he used to work or play outside coatless in the winter.

"That's what makes it feel so good to come in and get warm by the fire," he said then.

The mindset never left him. One of the times it surfaced in Southeast was during a season we were using separate boats – maybe we were depriving ourselves of each other's company in the daytime for rich reunions over hot grub in our tent at night.

"What an appetite!" I told him. "I'm finished and you're still going. I don't see how you can eat so much."

"You've been eating all day," he said "Big lunch, no doubt, and peanuts or something before and after."

"You mean you haven't?" I said.

"Not a single bite," he said, gnawing on a rib steak bone. "Being hungry makes food taste better."

"I see it must taste pretty good," I said.

"Satisfaction to be satisfying must be satisfaction that follows dissatisfaction," he said.

"Maybe we should avoid catching any fish for a few days," I said, "Look at how much fun catching some would be after that."

"Winter fishing would be the thing," said S. "When you get cold – that's still when it's greatest to get warm."

Spend Christmas handtrolling for salmon in Southeast? For pertinent comment, hark back to my Juneau newsbeat during the ice pileup on the Douglas shore. That was when fisheries official C. Howard Baltzo, as reported earlier, told me of foreseeing a big increase in winter commercial trolling in Alaska.

Now move forward to the fisheries of the 1990s. You find complex international and local considerations. Quotas apply to the chinook catch. There is winter chinook season and summer chinook season, and long closed seasons between. The quota is for the whole year. Winter comes first. Summer season lasts only until the quota is filled, and is brutally short.

Howard Baltzo was right. Winter fishing pressure has so increased that, as one troller puts it, "one of these years we will have to have separate summer and winter quotas or the total quota will be caught in the winter."

Conclusions?

1. The fishermen of old were pantywaists compared to today's.

2. The fishermen of today are ruining the joy of anticipating spring. What a loss! Before a body goes fishing he should have to squirm and fidget for awhile.

"There is," says S, "no joy in scratching unless you itch."

The Blahs

At eventide the chinooks bit and, on this first day of a new season, we boated four. Into a lonely cove familiar from past years we guided our two 20-footers and dropped the hooks.

An eagle eyed us from a rock. A crow flapped over and cocked his neck at us, but said nothing. Silence. Peace. Tranquillity.

"After a few weeks like this," I said in a newspaper column dispatched from Ketchikan to the lower 48 two days later, "you get so you can put up with people again."

This endearing approach to my readers apparently was received by them in the spirit sent, and no recriminations followed. But the comment was otherwise crooked. A "few weeks like this"? Uninterrupted peace and tranquillity rarely last that long.

On a previous year when I was newspapering in Juneau, a sad young man dragged himself into the office one dreary June day and told a tear-jerking tale, which we printed.

He was, or had wanted to be, a commercial fisherman. He was also broke and despondent. He had invested his total savings of $3000 in a small old boat and new fishing rods and reels. He had come the 300 miles from Ketchikan, fishing as he came. His luck had been rotten. He would fish no more. He would go back to work in the sawmill in Ketchikan. He would sell his whole fishing outfit here and now for whatever he could get for it.

He must have sold it, because a year later I encountered him not far from Ketchikan. He did not recognize me from the office in Juneau. He must have forcefully forgotten his wretchedness of then. He was equipped with another little old boat and several new rods and reels and he was fishing. Happily, it did appear.

Most switches between such highs and lows of spirit require no such passage of time. A day will do, or even hours or minutes. Fortunately, on a lows-highs scale of one to 10, Southeast normally sustains you within a point or two of seven. As you can see, a low is further down than a high is up.

Seldom can either a high or a low be traced clearly to a cause. If I was correct when I speculated a few pages back about why the deluge in Trollers Cove put me on a high, that high was the exception. And lows baffle you no less.

"We are gregarious," says a psychiatrist I have read. "We

need the company of our own kind." That's also what I said, a little less daintily, in my "put up with people" remark above. But both views dodge the specifics. You cannot analyze a low. They come and go as they please. They don't tell you why. You can survive poor fishing, ugly weather and fouled sparkplugs – and mire down in sunshine. "Nothing in the environment does it," says S. "Must come from your insides. Hormones."

Hormones. You secrete them at random. Then, in awhile, you produce counter-hormones, and the low lifts. Until it does, beware. A low is both omnipresent and untouchable. Eyeball to eyeball you cannot confront it. No system nor logic attends it. Southeast's green forests and snow-blue mountains waste their majesty on it. Indeed, by contrast they may aggravate it.

The malady has these parameters:

– It hits each of its subjects three times in a 120-day season and lasts one day each time. During 114 of the 120 days, therefore, neither you nor your partner is afflicted, and you skip along near level seven.

– It despises sympathy. If your cohort croons, "Everything will be all right," you will surely get worse.

– It will, per the numbers above, get to two of you simultaneously only at odds of 3/120 or once in 40 years.

We have not yet faced this frightful happenstance, but you may – and how to cope? Given the transient unpredictability of the hormones and the pitiful sights they will make of you, we suggest laughing at each other. At zero on the scale you can go nowhere but up.

Fooled by Fish

Fickle fish. Confounding fish. "You have to watch 'em," says Brother S. "Slip up and they don't let you forget it."

Example 1

My bones ached. My air mattress received me with a softness beyond eiderdown. "I've been working too hard," I murmured to Brother S. "The days are too long. "We haven't been catching much. Tomorrow morning I am going to sleep myself out."

And sleep I did.

When I came to, all the world was awake. Sunshine glowed through the tent walls, making a framework of the seams. Two crows were crabbing in a tree. A hummingbird buzzed and a squirrel chirped. Inside, the stove sizzled. In front of it, Brother S perched on a stool, brewing coffee. The aroma brought me full awake and I observed that he wore wet boots.

"You been fishing?" I asked.

"Yeah – I guess you'd call it that," he said.

"Catch any?"

"Six."

"What?"

"Six."

"Six?"

"Six. One goes 47 pounds. The six add up to 197."

"Six.... 47.... 197....?"

"The biggest one got away. About 55 pounds. Had him up to the boat three times, almost finished, when the hooks pulled out. Hung on another herring and this 47-pounder ate it."

That was the morning the fish had let me sleep.

Example 2

The day was overcast and ugly, black-clouded and foreboding, drab and gloomy, fishless and mirthless. Southeast's matchless scenery grew old as we panned along it from the boat, and we drifted into inattention. This was the condition two chinooks were waiting for. They capitalized not only on it but on the high-frequency hearing loss not rare among older folks like us.

We did land the two fish, but by no virtue of ours, as admitted in this post-excitement exchange:

Brother S: "Brother E, those poles must have been jumping up and down for five minutes. Why didn't we see them?"

Brother E: "Didn't hear the reels, either."

S: "Seldom do. You know that. When they turn fast the pitch is too high for us."

E: "But we did hear them, when the line was almost gone. I know – when the line goes out the spool gets smaller so it turns slower and the pitch gets lower and we hear it."

S: "Except ... the reel has to turn faster, not slower, when it gets smaller."

E: "Hmmm. I guess so. But it gets harder to pull against the drag, too, so it slowed the fish down. Matter of fact, they were stopping. That's when we heard them."

S: "Sure seems to be so."

Speaking of tuned-out ears, we occasionally fish near another old geezer who has the same problem we do. He solves it by fishing with his cocker spaniel. When the reel shrieks the dog barks. Dogs, as you know, can hear pitches higher than the most soprano of human ears. No fish ever fools that dog.

Example 3

The salmon were all on the surface. We had already landed three good-sized chinooks when S again leaned back with rod bowed. "This one's bigger," he said.

The fish made several runs, covering all directions, horizontal and vertical, but the rushes were short.

"May not be a salmon," I said.

"Salmon," he said. "What else?"

Not a good question. A salmon is a salmon because he is a salmon, not because he isn't something else. I peeled my eyes for a first glimpse of him.

But I began to doubt my doubts. S was sure, and S makes few mistakes.

"Salmon," he said once more, and continued the battle.

After further give and take, the fish ran toward the boat and under it, and came up on my side.

Clearly I saw the broad, brown, black-marked back of a sizable halibut. Halibut do not hit on the surface, but this one did.

Halibut were out of season. I hesitated. No, I would say nothing. Poor Brother S. He could make his own discovery.

But the deceitful halibut foiled that idea. On the next run he escaped, and I had to tell Brother S the truth, lest I inadvertently surprise him with it in print sometime.

Example 4

(Brother S got even with me. He wrote the following):

TOO MANY COHOS
By BROTHER S

What happened to Brother E and me went like this:

9:15 a.m. Nothing biting. Nice cloudy-bright morning, though. Calm water. Gentle tides. Beautiful world.

9:20 a.m. "Wup!" I say. "Wup!" E says. Two cohos on.

9:30 a.m. Both cohos in boat.

9:35 a.m. No more strikes. Still fishing, though. Four lines in water.

9:40 a.m. Fish on Pole One. Then on Two and Three. Then on all of them – "Four-bagger," says E.

We grab a pole apiece. The other two rattle in their holders.

My fish runs toward boat. With my help he leaps right inside. I put pole back in holder. The used bait dangles in water. I grab another pole and land fish on it.

E is still working on first one. His second pole is still bouncing. I pick it up and start cranking. Behind me, the one I left dangling sounds off. E has just got his fish in so he grabs that one. We boat them all.

"Something's cockeyed," says E. "Or I am. We had two fish and we caught four and now we've got seven. What's going on?"

Being a kind fellow, I tell him what happened.

"Whew!" he says. "Thank you!"

The Oddest Fish

"What a fighter this fish was!" the young fellow said, pointing at the chinook on the planks between his feet. "He's 19 pounds and I'd have sworn 50.... I'd hooked him in the tail."

The first time you heard that kind of story, you had your doubts. But you've since observed that chinooks smash into baitfish to disable them, and so get caught not only by the tail but by the back, belly, flanks, fins, and gill plates.

You've further noted that these foul-hooked fish, compared to "conventional" ones, bring to bear more power. And of these, the one attached exactly backwards outpulls all the others.

Correction. All but one that isn't attached at all.

Several times herein, while relating some experience, I have commented that only a few times a year, or even only once in several years, might you go through one like it.

Now for one that has happened exactly once in all my years.

The morning was clear and beautiful. I had shut off the motor and was drifting with an eddy. I had sunk my lines into a countercurrent below, and they angled back enough to indicate good bait action. No work was necessary. Peace prevailed. Relaxation was upon me.

The chinook was in relaxed mode too, apparently, basking in the sunshine. His dorsal fin stuck out of water and his tail waved lazily. He was on the other side of the eddy line, facing into the flow, tail to me. We were 20 feet apart and being slowly brought together.

I stared. Why would any fish lie there like that? *He must go 30 pounds,* I whispered to myself, and kept staring.

Under conditions which this fish did not fulfill, you will see salmon at or near the surface rather routinely – chinooks slashing through the baitfish or cohos pursuing individual, fleeing herring – sights exciting but not mystifying, being standard behavior. Other times, salmon will make darting passes at one that is on your line. And infrequently you will see them simply wandering around, cruising, without apparent purpose. What all these have in common is that they are moving. If the big chinook in the eddy were any more near motionless he could not have stayed erect.

I thought about casting a spinner in front of him. The motion would surely spook him. The range dropped to 10 feet....

My bow was beyond him. The sun illuminated him from one side and my white hull from the other. In my fishing seat at the starboard transom I was perfectly positioned.

Five feet. I was now beginning to believe I would finish this strange stalk successfully. I sneaked my hand out, poised to plunge it to a grip on his tail and in one motion heave him aboard.

It entered my head that he might be some way ailing. It further entered that the buyer would not know this.

But I turned noble – *No, you'll not sell him if he's sick or something.* Twelve inches. Now! My hand closed on firm flesh.

If that fish was sick, I would not wish to tangle with a well one. Luckily I can squeeze only about 200 pounds. At 300 I'd have lost my arm at the elbow. How that broad tail went through my grasp remains a mystery. My fingernails pierced my palm. The salmon simply disappeared, and I can enlighten you no further.

Fishermen's Semaphore, Etc.

Let me tell you about Fisherman's Semaphore, and also about the Shelton Brushpicker and the Niblack Liar.

Fisherman's Semaphore is a marine communications system. It is immune to electronic breakdown, there being no electronics. By manipulation of arms, hands, fingers, and other bodily parts, you exchange catch information over the water.

Fisherman's Semaphore originated in the lower 48 states. This was long ago. Fishermen were few, and all could be served. Accommodating all today would leave you too arm-weary to attend your gear. Consequently, in most places Fisherman's Semaphore is extinct.

Not so in Alaska. True, it has become threatened in recent years, and from crowded close-to-town sportfishing areas is largely gone. Outside these, though, you'll still send and receive intelligence efficiently. You need no training. Instinct suffices.

The basics are a show of fingers for numbers of fish and a spread of arms for size. If you have caught some big kings, three or four feet long, a spread of five feet is used, or as near five feet as a short-armed fellow can reach. For little fish, like cohos only two or three feet long, spread only a foot. Since all spreads are symbolic rather than actual, you don't mind exaggerating smallness for clarity. Say you have three nice kings and two cohos. You thrust three fingers skyward and spread hands wide; then two fingers and narrow. That's it.

That's it, that is, at quarters close enough for finger visibility. At long range, thrust your right arm aloft once for each fish and then transmit handspread size. If you've caught too many cohos or pinks to count, hoist both arms three times in quick succession. Kings? You never have too many to count.

Initiation of exchanges is more or less automatic, since both parties are sizing up each other on approach. The first step is to ask the question, "Fish?" This is done by an inquisitive tipping of the head and a tentative in-and-out spacing of the hands.

To transmit a no-fish reply, should you have none or see benefit in denying having any, hold both hands palm-up in front of you and cock your head sidewise while twisting and hunching your shoulders. For emphasis add a sneery, nasty expression and a sigh. These cannot be perceived at a distance but they help produce the right pose. For additional impact, if you have not caught a fish for several days, sag dolefully at the waist.

Be prepared for those times when you receive positive signals and in honesty can send only negative. Bear the trauma. Mercifully, you will soon be out of range. Besides, the guy may be less trustworthy than you. Or even you may wish to tell the truth to only one of two boats close enough to read you. Bosom friends develop nuances of style as unintelligible to outsiders as the antics of the third-base coach in the World Series. With experience you will develop a feel for false reports, good and bad. "It's been real hot here," I was told, back when I was still gullible, by a character who then departed, leaving me to bore holes in barren water.

Now about the Shelton Brushpicker and the Niblack Liar.

The Shelton Brushpicker got his name because he came from Shelton, Washington, and, like assorted other people in the western part of that Evergreen State, picked ornamental evergreen brush for a living part of the year. Another part he kelped for salmon in Southeast. The Niblack Liar got his name from me.

The Shelton Brushpicker was responsible for my audience with the Niblack Liar.

The Shelton Brushpicker by reputation was not a type to forego Fisherman's Semaphore and, instead, stop fishing and pull up alongside someone to chat in plain old oral language, and neither was I. But we did, on a warm June morning, off Caamano Point. I think the idea was more his than mine.

He aimed an arm down Clarence Strait toward Niblack Point. Said he: "Some special kings go past there this time of year. They're headed for Myers Chuck. Big ones. Powerful. Wider and heavier than most. Myers Chuck slugs, they are. I can't get down there today. Gotta go to town. You oughta try it."

Believe tips like that? But I did have to fish somewhere. I sunk my gear and trolled that direction.

Near Niblack Point one fisherman was working. In an hour I approached him. He had stayed in one little area.

This was the Niblack Liar, though I didn't know it then. But no matter. Time for Fisherman's Semaphore. I gave him the tipped-head, hand-spread "Fish?" question. He gave me the cocked-head, hunch-shouldered, flat-palmed "no fish" answer.

As already acknowledged, disclaiming good fishing is a legitimate form of deceit, even among the helpful people of Southeast. Sort of a game. There are other games. Radio codes, for example, say one thing and mean another. How come this guy stayed in one place? *Should have put glasses on him – seen what he was doing,* I thought.

I was 50 feet from him when his kicker quit. Hah! Before he could start it again I hurried over to help him (but mostly to have a look), and – Hah again! Under wet gunny sacks on his floorboard were four slug-sized bulges, 40-pound neighborhood.

"Fish?" I asked, in real oral language.

"Four," he said.

There are urges you must suppress. How would recrimination help? In my restraint I only asked, "are they deep or shallow?"

"Right on the bottom," he said.

Guided by his track record, I trolled away on the surface, and hadn't gone 100 feet when I hooked a 30-pounder, short and heavy, surely a Myers Chuck slug. The Shelton Brushpicker apparently was honest. But the Niblack Liar? He was as untruthful face to face as far off.

(This account was written in 1964. Now, in the 1990s, fishermen's semaphore has become largely a lost art, even in Alaska.)

10. THANKS FOR EVERYTHING

Fish, fishing, and fishers – they make a better place of this old planet and, to be sure, of Southeast. Before I can consider this journal complete I must offer a few tributes where I am sure they are due.

Awesome *Oncorhynchus*

Over the gunwale I got a glimpse of the salmon before he raced off again.

"No doubt about it," I told S. "That fish's tail is red. At least the top half of it is. Didn't you see it?"

"Didn't have time to look," he said. "Can feel him, though. He's powerful."

"He's not big," I said. "Twenty pounds, maybe. But red-tailed." I wiped my eyes. I had once seen a red-tailed sea gull – the work of some joker with a paint brush – but this was a fish....

In the landing net the puzzle ended. What we saw on that salmon was not paint, of course. It was bloody flesh. From his tail a foot forward, a fresh, raw wound extended. The bottom half of the tail was gone.

Sea lion? Seal? Dogfish? Who knows? What we do know is this: The genus *Oncorhynchus* – Pacific salmon – must be one of the Good Lord's most remarkable creations.

When my wife and I in 1958 bought the property on which we live, the final selling point was the salmon stream that crossed the 15 acres. I was exultant. Each fall now I'd be able to stand on my own real estate and cheer the returning run.

And I did. My newspaper column of Dec. 9, 1961, included these words of tribute:

> ...and when the salmon that hatch survive their fresh-water days and go to sea, they face constant new perils – the larger fish, the sea birds, the seals, the sea lions and that super predator, people. The salmon is pursued endlessly for food, pleasure and profit with bait, spoons, plugs, flashers, flies, nets and seines. Why, every fish that mysteriously follows his nose back to his home water to spawn must have logged a career of narrow escapes to make his blood run hot....

The evidence of predators, and especially of the super predator, presents itself repeatedly in Southeast salmon – wounds on their flanks, hooks in their jaws, rips where the hooks pulled out.

The classic embedded hook is a small one in a sizable fish – the big one that got away. Once I lost a rig in the morning and got it back from a 30-pound king in the afternoon.

Several times we have retrieved big power-troll hooks from the mouths of kings.

If you are impressed by the way these fish carry on with metal in their mouths – you haven't seen anything yet. Much more numerous are fish jaw-damaged where the hooks pulled out. We have taken many of these. One day, three we landed in succession were all hook-scarred.

All apparently had tangled with power trollers. One had lost part of his lower jaw, but healing was complete. The other two had new wounds, perhaps incurred the same day, or the same hour – and one had a new wound and an old, healed scar as well. On our lines, none of these could be distinguished from any "normal" fish.

Power trollers will tell you, to your increasing respect for your quarry, that their fishing gear does indeed fail much less often than does the fish's jaw. "And the biggest salmon are the ones you lose this way. Pull hard on a little fish and he comes. But pull hard on a big one and he can come apart. Rubber snubbers cushion the shock, but not always enough."

Brother S notes that king salmon – probably because they are the largest and longest-lived species – provide the most examples of the grand will to survive. Not all. Others of ours include a coho with scarred skin in place of his right eye and a humpy (pink) seemingly healthy despite an unbelievable jog in his back – at least half an inch of offset sideways between aft half and forward.

Once the salmon's physical achievements have adequately baffled you, move on to the realm of mysterious senses. To *Oncorhynchus* these are natural. To *Homo sapiens* they are unfathomable.

Consider the salmon's renowned ability to return from far seas to his personal spawning stream. The Atlantic salmon and all five Pacific species perform this feat, while researchers try to figure out how. As a salmon fan you're entitled to hope they never succeed.

In August 1968, venerable *National Geographic* magazine announced that the attractive smell theory of homing had been confirmed. Said a doctor-degreed source: "The fish smells its way from the coastline of the sea, tracking a familiar scent like a foxhound."

When researchers plugged a salmon's nostrils, the article said, he could not find his creek. Experimental proof!

Really? Why must the achievements of fish be decipherable in terms of the sensors of humans? Smell? What other function might nostrils enable? And what inviolable scent in the water was it, anyway, that was still the same in the dregs of autumn when the fish returned as it had been in the freshet of the long gone spring when he departed?

Sure enough, in November 1981 *Geographic* backed off. How the salmon gets home became "one of the great mysteries of biology" and "no one really knows."

In the September 1986 issue, odor regained ground: the salmon "navigate by smell upstream to where they were spawned."

But in a still more recent edition – July 1990 – compromise seems the stance: "The fish find their birthplace within a small margin of error, guided by the smell of the home stream and [italics mine:] *by other means that humans do not yet understand.*"

Man, those "other means"!

The article attributes the odors of home to "soils and plants along the rivers." The salmon is seldom foiled; he can detect a dilution of one part in a billion. *Billion.*

Forgive me for triple-barreled ignorance. 1. How do we know that soils and plants provide the odors he detects? 2. How does he know which of all the various contaminating parts-in-billions to track down? 3. Why, again, are these parts-in-billions forever unchanging?

And the "other means"? Among old postulations as to them are solar navigation and the detection of electrical currents as they vary with ocean currents. Soon, someone will suggest sustained dead-reckoning – the fish simply commits to his registers the directions and distances of life's travels and, when instinct says when, prints out the route back to square one. A multimillion-byte man-made computer with gyrocompass sensing could do this, so why not a superior salmon?

But wait – need sensing be involved at all? Why not simply *instinct*, that phenomenon by which *Homo sapiens* excuses what he can't explain? Isn't instinct why a salmon homes? Why not also how?

Having given up on the homing puzzle, tackle these:

1. You are moseying along at a slow troll, savoring the wild beauty of Southeast while also beginning to think it would be nice to be catching something when – what's that in the water? Herring. Not many, but enough. Your senses whip to attention. Should be a salmon here. Right! One bites. And you get him. But the question is – how, in all those cubic miles of water, did he *find* those herring?

2. I was moving over flat water in a little bay when I spotted some herring. I stopped the boat, rigged a new bait, cast it 50 feet port, saw a straight streak of motion across the surface, and was instantly fast to a fish. The bee-line made by his dorsal fin still showed. It began at least 100 feet off my starboard bow. The scenario was as if that fish had been lurking out there, watching the silly fisher in the boat messing with the bait. When the fisher pitched it overboard he was going to eat it. But how far and which way the fisher would pitch it he had no way of knowing. At least no way known to the fisher.

3. Brother S put me off on a house-sized rock so I could cast for cohos. I looked around, chose a direction for my first cast, made it, and realized that in backcast I had seen a salmon near the surface behind me. I also saw him head around the rock and go out of sight, and a second later I had him on my line – that same one, I don't doubt at all.

No, we don't think salmon can really see prey through rocks – they have some other way of knowing it is there! And we have since been gratified by word from researchers that fish can indeed detect activity from the far side of barriers.

Where there are no barriers, detection range increases. "I think," says Brother S, "that when you put a bait in the water every salmon within 100 yards knows that it is there.

"Most of them that come and look at it just go away. It either has to be doing just what they want it to do or they have to be ready to grab it no matter what it's doing. When they're all ready – that's when you get the big bites."

Pinpoint detection got demonstrated one day for a Ketchikan friend we had taken out with us. A very wild salmon on S' line was cavorting up, down, and around us, 50 feet or more in every direction – and another was heckling that one, darting around him on all sides.

"See if you can catch him," S said to our friend. "Throw your line out."

"Yes...b-but – what direction?"

"Any direction. Let it down. He'll find it."

He did. Then we had two very wild salmon.

On sizzling line or hot platter – and by all means in the interim between – the salmon must be treated with respect. Take a vow. Never will you so mishandle a fish you've caught that the buyer brands him a "No. 2," acceptable only at cut price. No. 2s, fishers say, never escape detection by the buyers, who "can smell them," methodically flipping fish after fish from cart to scale but laying back the flank of the lone soft-fleshed one and shunting him to No. 2.

Some No. 2s are inadvertent, like those valiantly living with wounds inflicted by predators. These will swell your esteem more than your coffers. Brother S took a champion among them, a stalwart whose power on the line gave us no clue that, in his side, a hole had been gouged to the depth of his backbone. At least a pound of flesh was gone. The rest of him weighed 33 pounds.

"What I don't understand," said S, "is how he ever got away right after that was done to him."

You could excuse No. 2 status for that fish. One inexcusable cause is belly-burn: acid-caused deterioration of the viscera, resulting when a fish is not promptly dressed, especially in hot weather. Another is flesh carelessly sliced by a cleaning knife. Never! No salmon should be so defiled.

The Honorable Herring

The salmon charged, the water boiled, and all the herring leaped.

"Whew!" said Brother S.

"Whew!" said I.

Salmon attacking baitfish – the sight, we've told you several times, turns us breathless. It always will.

We couldn't catch this salmon – sometimes you can't – but that's all right. It's his prey that I'm admiring right now.

Herring – in the wide world of this shapely and silvery little fish, we moochers are small potatoes. *Clupea harengus* and kin are distributed afar. The multi-ton winter haul that fills Alaska fish-buyers' plants with enough frozen bait to last all season is dwarfed by total world production, much of it for human consumption.

Smoked, pickled, canned, or otherwise, herring are nutritious and delicious, today and long ago. In about 1925 a friend of my father's took me fishing with the remark, "If we don't catch anything we'll eat the bait."

Locate a school of herring while you're mooching and your heartbeat reacts. You don't have to see salmon. The herring stir you by their sheer potential.

"Never leave a school of herring," a veteran fisherman convinced us in our youth. "Fish around those herring. Maybe no salmon will be biting.

"But fish there anyway, because, friend, that's where they'll be when they do decide to bite."

The only salmon bait superior to a herring, we contend, is a scent-oozing plug-cut or filet of herring.

Artificial lures – spoons, plugs, flies, and those oversized, heavy-pulling flashers yanking any of the lures behind them – all these catch fish sometimes.

"But the trouble with the artificials," says S, "is – which one and which color is right today? Herring are right all the time."

There are several ways to get herring. Our first source is the fish buyer, who sells them mass-frozen in pasteboard boxes. Box sizes have varied over the years. A 33-pounder is practical. You peel off individuals as they thaw. When they thaw at the same rate as the salmon are biting – that's just the way you had it figured.

If the fishing is slow and thawing exceeds use, slice a supply of filets and salt them down in a gallon can. They'll last all season. They'll freshen out on your line. They are tough – stay on the hook. And they catch fish.

When mass-frozen herring are too soft to stay put, as they sometimes are, you can use individually frozen sport bait, at a dear price, or catch your own.

Herring gillnets, herring rakes, and snag lines all work.

So does another idea, one that may have originated in the San Juan Islands of Washington State in the 1920s. An enterpriser there was supplying fishermen with bait herring from a source at first mysterious.

But the secret got out. Herring bite on bare hooks. He was catching them in deep water on a dozen shiny No. 8's spaced at six-inch intervals on a line.

Now, 70 years later, bare-hook jigging for herring remains a trick to remember in a pinch.

A sport-fishing acquaintance of ours jigs for his bait with a fly rod and single hook, with or without hackle.

"An eight- or nine-inch herring gives you a tussle on a fly rod," he says. "If I don't catch a salmon I've still had a day."

Occasionally a herring bites on the hook-point protruding from the filet you're offering the salmon, and you pull in your filet and the fresh new herring besides. And this happened once:

A salmon stripped my bait off, and a herring immediately hung himself on the bare hook. I was reeling him in, mentally making two spinners of him, when a 10-pound coho swallowed him.

I dressed that coho, fileted that herring, and caught a coho on one of the filets.

You won't often need to get bait from inside salmon, but the quirk will tempt you.

As for our resolve never to leave a school of herring, one test of it came during one of the clear-water days I spoke of earlier.

Mid-afternoon. We were passing a long, low point when we saw a hopeful sign: two dozen herring gulls perched on the rocks at water's edge, all facing out. "They're waiting for something," said S.

Never leave a school of herring? Never ignore a flock of waiting sea gulls, either. Join them in waiting for awhile. Find out what they are waiting for.

The clear water soon told us: herring, strung out endlessly in both directions. The problem: they were about eight feet down, too deep for the gulls.

"Millions of them," said I. "Billions," said S. Around this many herring, dozens of salmon could "decide to bite."

The gulls waited. A couple of eagles flapped down to wait with them. What they wait for is – must be – some signal that says "now!" to the life below.

When the signal sounds, we who have been there know, one of nature's great shows turns on. From the underside and the flanks of the herring the salmon attack, battering the baitfish silly and gobbling the quivering victims. The surface foams white with frantic herring and the gulls plunge at them with wild, exultant cries. It is a whirling circus.

We mooched and waited, mooched and waited. With the sea gulls and the eagles, we waited. Mooched and waited. And waited.

Not a nibble for three hours. Frustration set in. Around herring your metabolism zooms and you get drained fast.

But never leave a school of herring. We didn't. They left us. Sank deeper and disappeared. The sea gulls went away. And the eagles. And us. Night was near. With clear consciences we headed for our anchorage.

It took no conference to bring us back at dawn. In print I recorded the result thus:

> Along this rocky shore there was a mile of herring – leaping, squirming, making a noise like rain as they churned the surface. This was in the mild places. In the wild ones, flocks of sea gulls, supplemented by a few eagles, were screaming, swooping, and scooping. Everything, everywhere, was in motion.
>
> We lowered our baits in stop-and-go fashion into the writhing sea and for two hours had few unfilled minutes – or seconds. We had two troubles. Cohos and small kings often beat the big fish to the bait, and the big ones took too long to land. On the floorboards of our messed-up boat we dumped 17 salmon: seven cohos to eight pounds, three humpies, two small kings under eight pounds and five large kings to 41 pounds. And for good measure I added the night's dinner – a seven-pound steelhead.

After the big two hours the herring faded away and the fishing became what you could call normal.

For several days thereafter we worked around smaller herring schools without finding anything like the big one.

But never leave any of them. When you find one, remember where it was. At other times, scout them out, by checking eddies, currents, and gatherings of drift, and by watching diving birds – not to mention gulls waiting on rocks. And again, when you find the herring – never leave them.

Fishing's Gifts to Me

Fishing, as you know, tones down little troubles.

And three times it has rescued me from bigger ones.

Southeast salmon mooching was bound to be the climax.

In the 1920s when Brother S and I were boys, the forest-shaded waters of Puget Sound still offered what Southeast gave us later. There was nothing like Saturday in a rowboat. Unchallenged for space, we stalked perch or rockfish and, for alternative fascination, beached the boat and explored the woods.

One day S noticed a picture in a magazine depicting "great angling" somewhere back East. Shoulder to shoulder the anglers stood, backs to camera, rods slanted away in regimented alignment for clearance.

This was fishing? It never struck us then that we would see the day in our own West when clients fished en masse over the rails of charter boats 50 feet long.

But the day came. And, I am told, these people even enjoy themselves. But they have never been *fishing.*

Then again, the merits of fishing are varied and numerous. Let me get to the three rescues.

Rescue One

The time was 1950 and the place the Journalism School at the University of Washington.

I was too old to be there, having graduated from high school 16 years earlier. My record in the meantime had been totally unspectacular, and was in no way enhanced when a perforated eardrum exempted me from military service during World War II. My rationale for enrolling anyway was that my colleagues would include students of ages approaching mine, veterans capitalizing on the GI Bill.

Enrolling was one thing, performing another. The old college saw of "publish or perish" could hardly be more true than for journalists, and our class projects included trying to sell articles to magazines. Articles about what? While my worldly classmates wrote of McCarthyism, civil rights, and foreign governments and contemplated careers in international affairs, I wrote of tackle, trout, and salmon and contemplated retreat from higher education.

But my professors – ah! Several of them were fishermen. To them I was a welcome diversion in a mind-wearying world of amateur statesmanship. They backed me solidly. In a piece about fishing in the Skokomish River I reported a catch that included an unbelievable four-pound cutthroat trout. My professor happily appraised the piece as ready for trying on an editor. He also red-penned a note in the margin alongside the "unbelievable" four-pounder from the Skokomish. "Don't apologize. I got a five-pounder in the Stillaguamish."

About this time the fascinating technique of mooching was sweeping Puget Sound and I sold a piece about it to *Field and Stream* magazine. I did not keep this achievement secret. My professors loved me and my colleagues pretended they did.

So I graduated. Then I went fishing. Spent the summer mooching commercially or "kelping" out of Neah Bay at the State of Washington's northwest tip.

It was a great summer. Probing the kelp borders and the recesses of crooked, rocky shorelines with two or three outfits at once was a consummate pursuit. I saw one way to improve upon it: do it in Southeast Alaska, where the moochers were few, the salmon large and plentiful, and green islands with rocky shores everywhere and endless.

My diploma got me a reporting job on *Daily Alaska Empire* in Juneau – a job that was also a means to an end. I could finish my newspapering at mid-afternoon and mooch until near midnight. This was the life. The only people I envied were the school teachers. They could fish for three months unadulterated.

Rescue Two

The Juneau winters and a circulatory deficiency ruined my Shangri-La. In near-zero winter weather my hands, feet, nose, and ears froze despite the best clothing available. So now what? I went job-seeking at *The Sun* in the milder climate of my home town of Bremerton, Washington.

"We're fully staffed," a secretary told me. "You can talk to the editor if you want to, though. He makes the decisions."

The editor was sympathetic but not encouraging – until I got through to him with my background. Then he perked up.

"Alaska....fishing.... Everybody is interested in Alaska and fishing these days," he said.... "Let me see...."

And I swung a deal (he thought he swung it) whereby I became a beat reporter most of the year and Alaska "mooching correspondent" in the summer. How's that for a bailout?

Rescue Three

My *Sun* career got stymied in 1966 by a hearing loss. I began fearing misquoting the mayor at a meeting or the witness at a murder trial. Whew!... And besides, by this time, any disrepute I incurred would fall also on a dear wife and twin daughters.

A nearby federal weapons establishment was advertising for an editor for its scientific and engineering reports. What's to lose? I applied. The head man (GS-15, to federal pay-grade watchers) interviewed me.

"I've read your Alaska fishing stories in the paper," he said. "Great fishing. What do you know about engineering?"

I said, "Uh....well...."

He said, "The other applicants haven't known much, either." He added that he was looking for someone who could simplify the complicated verbiage his learned engineers were producing.

"Simple? I can be simple," I said.

I was seated in the new job while still drawing vacation pay from the old one.

Rescue Three and a Half

Getting the position proved easier than filling it. The learned engineers did not readily accept my simplifications.

But fishing buffered the battlefield. Many of the reports I handled covered tests of torpedoes fired from submarines on a test range in Hood Canal,* waters also occupied by salmon and frequented by range engineers fishing while off duty. When debate over unreadable verbiage raged, I came up with bait-cutting tips and hot salmon locations.

My Southeast summers did not fare too badly, either, vacations being generous in the federal service. Meanwhile time marched on to retirement, and since then, in recognition of the contribution of fishing to me, I've returned the favor with additional attention to fishing.

---

*One submarine is said to have surfaced at headquarters towing 200 yards of fishing line. Some story that fisherman must have told!

Brother S

Some fishermen are born and some are made.

A born fisherman, I say, has no basis for boasting. He has achieved nothing not built into him. A made fisherman has.

I may have been somewhat born. S was surely made.

In the beginning he was neither. He had no imagination, no flair for improvisation. He fished mechanically. He caught fish because there were plenty to catch.

This I say now; assuredly not then. Until well into adulthood I could never as much as have thought it. If Brother S fished a certain way, it was the right way. I fished as he said....

My story of Brother S entails three Big Switches.

Brother S (for Svend, if you've forgotten) was born August 1, 1909, on a waterfront farm on Puget Sound, near Bremerton, Washington. He was the third of six children of Chris and Valborg Jensen, immigrants from Denmark.

Seven years later I arrived. I probably began attaching to S about three days after that.

When I look back now from the attitudes of today – how did he so willingly tolerate and educate so much younger a sibling as I? From his vast and growing store of lore, he briefed me on the wonders of the beach and the waters in front of the farm and the woods on the hills behind....

When I was 21, Brother S got married.

This event was not as bad as it sounds at first. In fact it did not quite rate as one of the Big Switches. Although Brother S and his lady raised three children, S never did abandon me.... And we had our Southeast years.

The first Big Switch came in the 1940s. While Brother S was still mechanically towing plugs and spoons through salmon water, I started mooching. And catching fish. And coaching *him*. Imagine it! Now I was the instructor and he the pupil. The first Big Switch was history.

Brother S was a poor student. Conservatism held him back. And in our early Southeast seasons of the 1950s, when I had progressed to plug-cuts and filets, he still favored plain whole herring. Furthermore, the herring had to be uniform seven- to eight-inchers, the one size he'd learned to rig so they spun exactly to the specifications he doggedly made gospel....

Late June 1964. Second Big Switch.

By this time S had retired from his job as a supervisory rigger in the naval shipyard at Bremerton and could spend whole seasons in Southeast. I was still on the *Sun* staff, with a set schedule for return from field to newsroom.

And so, as on at least two other occasions when momentous events took place in Southeast, I was not there.

I hold no grudge against the run of kings which, two days after my departure, crowded into Vallenar Bay. What pains me is that I missed out on the maturation of my student, Brother S.

Those chinooks, he told me, "were lining up to bite."

They arrived at mid afternoon. On previous days there had been none. Although the whole 40-pound block of miscellaneous bait herring we had purchased before I left had completely thawed, the only ones S had used were the few of the size that suited him. The "miscellaneous" rest were monsters a foot long. So – S had no bait.

What to do now? Run back to the floats, drive to the fish plant in Ketchikan, get the bait and then make all the reverse legs, while all those kings waited impatiently in line?

S told me about this in a letter. I wished I were fileting those jumbo herring (big ones yield tough spinners that stay on the hook) but instead I was 700 miles away, covering some dreary court case or some cow stuck in a well.

But the letter got better. Brother S' straitjacketed rules, you see, covered not only procedures but equipment maintenance, and "knives are useless unless you keep them sharp."

So his knife was sharp when, drawing on some latent emergency gene neither he nor I knew he had, he laid one of those huge herring on the seat beside him and, with a single jiggling pass of the blade, sliced off a neat tapered filet six inches long.

"I always wondered why you spent so much time at it," he told me later, referring to my system – indeed, the system then standard – which required several tedious preliminary knife cuts on the herring.

In Vallenar Bay that June day, Brother S sunk his first cut spinner over the side. In water 50 feet deep, it stopped in 15. And the chinook that had stopped it took off.

"After that," said S, "I had fish on as fast as I could cut bait. They were big and they were everywhere. Once I got two on at the same time and they went better than 50 pounds apiece. They ran opposite directions and it is an absolute fact that they both jumped at once, a hundred yards out on each side."

It may be occurring to you that looking two directions at two fish at one time is a neat trick. But since Brother S said it, it must be so.

Both fish, he said, shook the hooks, "and I lost another after 15 minutes, but I still got six before dark."

It made no difference that those lined-up kings would have hit anything. Brother S was a spinner-cutter from then on.

To his original emergency design he made one modification: he trimmed the sometimes ragged belly of the spinner on a cutting board. By the end of his sharp-knife career, he had cut so many spinners that the board was hollowed out like a butcher's block, a quarter inch deep.

His success with the lined-up kings of Vallenar Bay turned him adventuresome and he experimented with spinners of all designs, and kept track of the results.

"Size doesn't much matter," he said. "Neither does shape. Tight spin is the thing. You get that by putting the front hook in the right place. What's right? Doesn't matter. If you don't get tight spin, move it until you do. Cutting spinners takes no time and neither does rigging them. Just hook them so they spin tight."

And, he said, "Fish hard. Believe your bait will catch fish. Stay on the job. And keep your knife sharp."...

Now, let me make this as easy as possible on myself....

Where events in this book carry no date, none is needed. What happened could have happened any time in the 40-plus years the book covers, and still can happen. The basic elements of the Southeast mooching life have not changed. Dates are immaterial.

After 1984, however, one difference arises. Dates may appear, but Brother S does not.

A successor* now occupies Brother S' seat near the fishbox.

Nowhere else in the book do I reveal this. I refer to him always in the present tense, as if he still were with us.

During his shipyard service Brother S had breathed asbestos. Lung cancer resulted (he never smoked) and he succumbed

---

*Bill Clark, a retired junior high school band director with diverse talents unrelated to that one, including adaptability to mooching.

in December 1984. It was the third and final Big Switch. I was again in charge.

And I do keep my knife sharp.

EXPERT IN ACTION: Here is Brother S at work cutting bait, along with the finished product.

A Place Made for Mooching

When learned geologists view Southeast Alaska they lose their scholarly cool and indulge in superlatives. Southeast is "wildly upheaved," "grossly buckled," and "fused by molten lava into preposterous, enormous shapes." Everything is great. The Coast Range is "a great barrier to the great Pacific." It causes "great precipitation." Because of the high latitude and high altitude, this comes mostly in "great snowfalls." These great snowfalls have compacted into "great glaciers," and the great glaciers have "carved an amazing jumble of peaks, rounded shoulders, and tremendous furrows."...

You do not have to be a learned geologist to see that these (great) processes have been all to the good. They left half the "amazing jumble" above sea level and half below.

The half above brought forth the nation's greatest national forest, the Tongass, 16 million acres. The half below begot the world's greatest salmon runs, including the greatest king ever known to have swum anywhere or anytime, 126 great pounds. No great wonder, as Brother S says, Southeast is "a place made for mooching."

One day in October in the Puget Sound country of Washington State Brother S stopped by my house to say he had been "accosted" by a reader of the newspaper columns I'd sent back from Southeast during the summer.

"The fellow wanted to know," said S, "if he could float-fish for chinooks from an inner tube, like for trout in a beaver pond."

"Whew!" I said. But I knew the reason for the question. At least one of my columns had expressed fascination with the "made for mooching" aspect of the country. We had, I said, been "catching kings up to 50 pounds in water which, being among rocks and behind headlands, is perfectly flat."

But hardly did I suggest doing it from an inner tube.

"Well, maybe you could," Brother S said he told the man, "but you'd need an outboard motor to bring you back from where the fish and the tide might take you."

We still preferred boats. Ah, but there's nothing like a boat.

So about boats – what size will best serve a Southeast moocher? Hey, if you could remain always among rocks behind headlands, you wouldn't need much more than the inner tube. A pram? Now, there's some maneuverability and excitement for you. But – too bad – you can't live (comfortably) aboard a pram.

Fishing from a big live-aboard cruiser is not the answer, either – too unwieldy. But how about a mother-ship – a 50-footer to live on in a sheltered and secluded retreat near the fishing grounds and six little boats to fish from?... Availing yourself of the cruiser might not be too tough but...how to find the right six mates?...Those like I've had are in limited supply.

A boat both big enough and small enough is about a 20-footer.... Sleeps two, has galley and head, cruises at 20 mph with a 100-horse outboard, and has 15-horse trolling/backup power....

* * * * * * *

Next question: one such boat or two?

With two you can sample more territory, have fun keeping in touch by radio, and assist each other in case of breakdown.

The advantages end there. Going in one boat eliminates the expense of transporting, operating, and maintaining the second one... also saves on skippering time. ... Besides, when the fish are biting, two people can handle as much gear from one boat as from two, or nearly so. ...

Final word: in a season you can catch 25 percent more fish from two boats than from one... and use up the 25 percent on extra expenses. Take your choice.

* * * * * * *

Surprise: Avoid lightweight fishing rods. The stiffest of poles will give you all the shock-absorption you need, if you keep the tip up. Try this: hang a two-pound weight on your nine-foot rod and see if you can lift it with one hand from the handle. You'll be amazed. ... A rod tip that is pointing at the fish is not part of the rod but of the line. ...A stout rod gives you more control at netting time, too. ... The disadvantage is weight... a shorter rod can be lighter... so... what's the answer? another take-your-choice.

* * * * * * *

Second surprise: Leaders should be heavier than lines. It's the leader that takes the beating. ... Do heavy leaders scare fish? Or impede bait action? Commercial trollers use hundred-plus pounders. ... A dog's brain is larger than a fish's, and a dog will grab a steak hung on a rope. ... But light line? You bet. Gives you more depth and better feel. ... Twenty-pound line with 30-pound leader works well. ... Watch for flaws in both line and leader. Little nicks become big ones under strain.

Admittedly, if you snag bottom and the heavy leader holds, you can lose hooks, leader, sinker, swivels – the works. ... But at a cost of only a few bucks.... Just carry plenty of extras.

* * * * * * *

Third surprise: You needn't buy any of the high-priced "better than any other" line. Stuff costing a fraction as much has given us no woe.... Indeed, it may be stiffer and tougher than premium grade.... If you're uncertain, tie it to a tree and test it.

The above bits of advice bring us near the end of what I have aimed to say. What I must add is brief.

Here and there throughout these pages you have been told of our awe on encountering examples of nature's wonders. Their variety is endless. You don't have to be a Southeast moocher to meet up with these marvels but the daily exposure does pile up the for-instances. Here, selected from many, are three final ones:

* * * * * * *

1. On a sunny morning we were trolling lazily near a brushy shore when a hummingbird appeared, heading landward from outside. He whizzed across our stern, stopped on his target limb, and perched there, instantly motionless.

"I don't see how a bird does that," said S. "He was going full speed until he got to exactly the right place at exactly the right time, with his feet exactly on that branch. How did he even see where he was going? How do birds do that?"

* * * * * * *

2. All night the gale howled above us as we rode at anchor, snug in our little nook. The planet itself seemed shaking. "It's a real Southeaster," said S.

In the morning the wind subsided and we emerged to inspect the outer world. A long, dark object was afloat in Clarence Strait. We approached. It was a 100-foot length of green hemlock tree, two feet in diameter where it had cracked off. "The roots held, so the tree broke," I said.

So did others, before and since. Here and there in Southeast you will see the shattered stumps, short and tall, left behind by doomed trees like that hemlock.

We moved shoreward. Projecting toward us was the top of a downed cedar. We beached the boat. The bole of the uprooted tree was three feet through. The hole where the roots had been was hardly a hole at all, not a foot deep. The bottom of it was bare bedrock, almost flat.

"There's nothing there for the roots to hold onto," said S. "But hold they did, for maybe 200 years. How do they do that?"

* * * * * * *

3. The final thing I must express puzzlement about as I wrap up this written labor of love is – what urge is it that, in Southeast or elsewhere, makes a person go fishing or hunting?

He'll give you reasons. The outing, not the bag, is the thing. He likes the "sport," likes to "get out." Likes the fresh air, likes the campfires. Fine. But he likes all these things more if he catches or bags something, and the more the better.

He lauds his quarry and its rare elusiveness. He thinks its intelligence must surpass that of most men or women. But he lets you understand that he is able to outwit it from time to time.

Don't get me wrong. No way would I say anything against a deer, pheasant, or fish. I only think we should appreciate them for what they are and quit trying to understand. Why go fishing? Don't wonder. Just go....

FINISH A BOOK WITH A SUNSET? No, it's a sunrise: 3 a.m. in the middle *of June – surely the dawn of another day to be remembered.*